Nonlethal Weapons

Nonlethal Weapons

WAR WITHOUT DEATH

David A. Morehouse

PRAEGER

Westport, Connecticut
London

Library of Congress Cataloging-in-Publication Data

Morehouse, David, 1954–
Nonlethal weapons : war without death / David A. Morehouse.
p. cm.
Includes bibliographical references and index.
ISBN 0–275–95170–7 (alk. paper)
1. Nonlethal weapons. I. Title.
U795.M67 1996
355.02—dc20 96–16271

British Library Cataloguing in Publication Data is available.

Library of Congress Catalog Card Number: 96–16271
ISBN: 0–275–95170–7

First published in 1996

Praeger Publishers, 88 Post Road West, Westport, CT 06881
An imprint of Greenwood Publishing Group, Inc.

Printed in the United States of America

The paper used in this book complies with the Permanent Paper Standard issued by the National Information Standards Organization (Z39.48–1984).

10 9 8 7 6 5 4 3 2 1

To

Michael, Mariah, and Danielle

The starting point for a better world is the belief that it is possible.

Norman Cousins, *The Human Adventure*

Contents

Figures and Tables

FIGURES

TABLES

Acknowledgments

Whatever the weaknesses of this book, they would have been far greater were it not for the help of Colonel Jim Dubik. His invaluable counsel in the early stages of development and support of my philosophical beliefs even when others worked very hard to stop the project will always be remembered. To Jim Garrison, Ph.D., president of the Mikhail Gorbachev Foundation, go my thanks for believing in the project and for the opportunity to present its theory at the first annual State of the World Forum in 1995. To Senator Alan Cranston (D-California) go my thanks for his kind words of encouragement and his wisdom.

I am extremely grateful for the sustained support and friendship of my literary agent, Ms. Sandra Martin, who believed in me and this work for many years; she will never be forgotten. My thanks go to my dear friend, Jim Marrs, author of the *New York Times* bestseller *Crossfire: The Plot That Killed Kennedy*. Jim's counsel and encouragement through the years is appreciated. I thank my parents, whose love and example gave me the courage to believe that truth is the desired objective of all humanity. Most important, I wish to thank my wife, Debbie, who stood by me through sixteen years of military service, encouraging me in the face of adversity to do what is right, regardless of the consequences.

This book is dedicated to my children, Michael, Mariah, and Danielle, for whom I pray the world becomes a better place.

Nonlethal Weapons

Prologue

> As an Army officer I debated many times about whether violence is part of the human makeup or a learned behavior. . . . I can only say . . . that I remember clearly a time when I knew nothing of war and violence, and later how I had worked hard to learn to become capable of it.
>
> Major David Morehouse, Ph.D.

A PLACE TO BEGIN

Anger is genetic; war and violence are not. No matter how much we would like that fact to change, it will not. We come into this life expressing fear and anger even as we draw our first breath. Clinched fists, tears, and bellowing lungs are our first declarations of life. This is how we get what we want at that age: basic things like food and drink.

In the years that follow, we learn to transform our anger into various forms of violence. This transformation from fist to weapon is the genesis of war. It is not our fault that we make this transformation; our environment is saturated with concepts for the application of force. From the first colorful images we see in the powerful media, we learn that weapons are a means by which one person forces another to submit. This is the nature of war—one that will never change.

The task at hand is to change the way in which wars are fought, which is where the concept of nonlethality dawns. Humanity will always be forced to confront its anger. There will always be the straight-line cause and effect of humanism. Paraphrasing Eric Hoffer's "True Believer," circa 1930–1950, there will always be distrust, hatred, greed, desire, retaliation, heritage, and ignorance manifesting itself in our lives. However, how we confront our anger will make the difference between taking an evolutionary step forward or backward.

We are at a crossroads in time. The third millennium approaches, bringing with it a new era of thought and philosophy. The excitement of this new beginning is all around us, contagious and unavoidable. People on every front are gearing themselves up for change, and welcoming it. Global political leadership speaks in terms of a new world order, with new approaches to solving human crises and finding a way to live together in peace and harmony. Virtually every aspect of our existence is touched by this intervening philosophy; from science and technology to spiritualism, leadership, communications, economics, the arts, media, education, and politics, all are being altered by this drift toward transformation. It is time for the science and politics of defense to address the trends of the rest of humanity and seek more peaceful solutions to global contention.

How do we address these trends? It is clear that we do it alone. It is incumbent on us to set in place for our children the tools of this transformation. Our generation learned only more lethal methods of taking life, permitting us to kill on an ever-increasing scale of precision and volume. Moreover, current trends in military technology are digitizing the battlefield, thus further dehumanizing it.

This dehumanization, in turn, entices military and political leadership to abandon diplomacy and quickly ascend through the levels of power in order to declare total war. Their fear is to be accused of gradualism; therefore, they strike out with what they have—a lethal arsenal designed to destroy and thereby control—nothing more or less. The result is carnage, a two-edged sword cutting both ways, with human sacrifice on either side. Philosophically speaking, our generation's mission is to offer the future inhabitants of this planet an alternative to that inevitable, rehearsed destiny. Specifically, we need to develop a balanced force structure that offers leadership more balanced options. Nonlethal weapons and technology offer that alternative through the destruction of war machines, not human life; in essence, by allowing control without destruction. Clearly, we need a new silver bullet!

SOME BACKGROUND

Nonlethal technology offers the user an even broader inventory of weapons than is available to the user of conventional "lethal" technologies. Let's face it—the conventional arsenal either burns, explodes, or implodes, all with the same end result; the munitions merely come in different packages with varying delivery systems. We still bring the same things to the battlefield that we brought 130 years ago; they are just more lethal and accurate. Nonlethal weapons and technologies offer a wide range of options and varying outcomes. Many, if not all, of the current delivery systems can remain; only the end result and its effect on human life will change.

Given the dearth of attention that has been paid to the concept, the spectrum of nonlethal options is astounding. There are far more ways not to kill while ending conflict than there are to kill. The former run the gamut from calmative agents that make people lie down and sleep to munitions that can hold a battle tank in its tracks. The spectrum of choices is tenfold that of lethal means.

These are not futuristic weapons conjured up only in the imagination of some science fiction writer—they are real and available. They are, in many cases, what the U.S. Department of Defense might call "off-the-shelf" technologies, with some having been in existence for 50 to 60 years or more. Now, through the pioneering efforts of organizations such as the Mikhail Gorbachev Foundation's first annual State of the World Forum and others, nonlethality is being discussed in roundtable formats by the world's most powerful political and military leaders, opening a new era of understanding. The very idea that we can show present and future generations is a new road to the resolution of differences is the epitome of philosophic wisdom in practice.

OVERVIEW

In the post–Cold War world, the "superpowers" will seek to achieve their security objectives at reduced costs, with the fewest casualties, and in a manner most acceptable to the global public. Furthermore, this world opinion, aided by an increasing global media system, will play a significant role in determining the level of force that any nation can commit.

The international and domestic stages have changed. Recognizing this, the sensitivity of policy makers must change as well. In the future, the world must seek alternatives to conventional lethal technologies in an effort to maintain the moral high ground, at best, or to keep world opinion in abeyance, at least. Nonlethal technologies offer this option to international command authorities across the spectrum, whether in peace or war.

THE PHILOSOPHY OF NONLETHAL WARFARE

This book is intended as the first transitional record of a revolutionary method of war, using nonlethal technologies. For over 70 years, scientists and researchers worldwide have been developing, testing, and proving that humankind can create weapons that are antimatériel rather than antipersonnel. This landmark topic will be explored in full, using fictional scenarios to illustrate the application of existing classified and open-source, nonlethal technologies.

Since 1991, various collections of Defense Department working groups and individuals have been employed in the study of nonlethal weapons technology and its application to the new defense policy. Surprisingly, however, since that time there has been no significant progress made toward this endeavor by the Defense Department, its satellite focus groups, or its hired research departments like Los Alamos National Labs.

Can it be that the global defense industry, which amounts to $900 billion annually, is too powerful for politicians and defenders to influence? Or could it be that the concept itself is so convoluted and misunderstood that the world's leaders are incapable of codifying this multidimensional subject? Could it be that much of the real effort has been drowned in the politicizing of the concept rather than the honest promotion of its possibilities? This book does not attempt to drag the reader through the argument; instead, it provides a simple explanation for why nations build weapons and forces, and describes sample alternative, nonlethal technologies to augment the existing global arsenals.

While I believe that the preceding reasons for the stagnation of these technologies are true, I will leave the analysis and verdict to the reader. I intend only to describe the capabilities and applications of the technologies, the fight for their implementation must be a collective effort on many political and economic fronts, and is certainly not the focus of a work such as this. Draw your own conclusions, and take the steps you feel necessary. This book is meant only to establish a common point of departure for an expanded discussion.

The U.S. defense base includes existing nonlethal technologies designed to disable or destroy an enemy's war-fighting capability without causing death, excessive property damage, or widespread environmental destruction. Some of the technologies are hybrids—that is, a mix of the lethal and nonlethal methods. They are sometimes called "soft-kill," because they permanently blind or otherwise incapacitate enemy forces but do not kill. Others are revolutionary methods and systems that do not harm any form of life. Unfortunately, it is the hybrids on which the defense industry and the Department of Defense focus.

In a recent symposium sponsored by the U.S. Army Infantry School–Dismounted Warfighting Battle Lab, several advanced technologies of lethal and nonlethal categories were presented to a group of military officers. The officers listened intently to a somber illustration of a directed-energy weapon (a laser) designed to blind enemy soldiers. Interestingly, however, this technology was introduced as nonlethal; it does not kill but only maims.

One of the officers in the group raised his hand to say, "As a commander, in battle, I can say with complete certainty that I would much rather roll over a thousand dead bodies than contend with a thousand blinded prisoners." These words are not harsh from a warrior's perspective. Why, then, spend time and money developing a weapon to maim soldiers and then muddy the waters by calling it a nonlethal technology? In this volume, such hybrid technologies will be identified, categorized, and discussed.

The "true" nonlethal technologies are destined to alter the course of history; they do not maim—they preserve life. They offer humankind an evolutionary escape hatch from what is otherwise predicted to be a continuing cycle of ever-increasing lethality. However, these technologies are being held in abeyance—sitting on shelves—while the global superpowers strengthen their powerful deterrence capability with expensive, high-technology, "smart weapons" of the lethal category.

As leaders in the most powerful societies on the planet, we have an obligation to expand the role of nonlethal technologies in support of, and in some cases, in lieu of, lethal technologies. Humankind has lost its patriotic fervor for the waging of lethal wars. Any victory at the cost of life is a limited victory.

Critical to our understanding of the significance and role of nonlethal technologies is the need to grasp the magnitude of destruction that we have brought on each other. Totalitarian genocide has killed millions, and additional millions have been tortured, maimed, and displaced. As concerned global leaders, we have it in our grasp to end much of the devastation of wars around the world if we will only shift our defense efforts toward nonlethal technologies.

This shift will require a whole new way of thinking about defense and deterrence. Nonlethality is a revolutionary concept that can guide the international community into realizing a new world order. When nonlethal technologies replace the old weapons of destruction, diplomacy will take its rightful place as the supreme method of conflict resolution.

Chapter 1

To Kill: The Chosen Path of Humanity

> The acme of skill is not winning a hundred victories in a hundred battles, but to subdue the armies of the enemy without fighting. Thus, those skilled in war subdue the enemy's army without battle. They capture his cities without assaulting them and overthrow his state without protracted operations.
>
> Sun Tzu, *The Art of War*[1]

INTRODUCTION

Over 170 million innocent human beings have perished in war and at the hands of war makers in this century alone. Of that figure, a recently projected 80 million were, in essence, summarily executed for not acquiescing to the belief structure of the war maker. Think of it—80 million innocent human beings put to death! These were not warriors: these were political figures, doctors, lawyers, professors, housewives, schoolchildren, factory workers, and so on. They did not take up arms against the belligerent; they simply lived their lives and became caught up in the fire storm and carnage of war.

As human beings, we have spent the better part of this century designing, manufacturing, and perfecting weapons of mass destruction; of particular note is the global nuclear arsenal. World powers have spent several recent decades negotiating various controls and limitations for the strategic employment of these arsenals. Within the last decade or so, much effort has been focused on the ultimate abolition of these weapons. While undertaking the destruction of the nuclear arsenal is noble, it is futile as well. If, this very minute, it were agreed on in the council chambers of every power in possession of these weapons to assure their mutual destruction and outlaw their further use, it would still serve little or no purpose. The technology would continue to exist in the minds of humans—and the desire to employ its destructive capability would live on in their hearts. This is an undeniable fact. Therefore, it can be assumed that human nature will never change; that there will always be hatred, aggression, coveting, racism, poverty, famine, and all the deadly forces that drive humankind into a warring frenzy. If human nature will not change, then the nature of war will not change, but only the way in which wars are fought.

The current trend is to develop weapons of increasing lethality and precision, the result being the advent of a multibillion-dollar, global defense industry. As this industry grows unregulated and essentially unwatched by the global community, the "old guard" leadership trifles over the issue of mutual nuclear disarmament. We spend years in negotiations that result literally in the unscrewing of an insignificant number of warheads from delivery vehicles; and then we pat ourselves on the back as though we had made a difference. The nuclear arsenal is certainly a threat—but one so frightening that, with the exception of a few rogue states that might attempt to develop a nuclear terrorist threat, the world is virtually assured of the continuation of mutual deterrence based on the reality of mutual annihilation. Some may think this philosophy idealistic and sophomoric, while others may think it insane and misguided; but I suggest that it is reality. While we turn our focus on a nuclear arsenal that presents relatively little threat, a defense industry grows at an alarming rate, peddling weapons and technologies across the globe. For many nations, the chief exports are now weapons technologies. Instead of food, medicine, agricultural machinery, clothing and textiles, education, communications, and other productive technologies, they export and import weapons and technologies for defense and offense.

Humanity must find another path. The defense industry cannot be permitted to continue as it is currently growing. Of the 170 million casualties of this century, a mere handful—roughly less than 200,000—have perished as a result of nuclear weapons. However, because of its potential, the nuclear arsenal remains our principal focus, while we have let a conventional arsenal explode beyond all expectations. The arsenal that took the extreme majority of life expands unchecked and unlimited; and conventional weapons now approach nuclear technologies in their ability to kill in great numbers. Remember, precision and "smart" munitions are only one small, and recently visible, aspect of the battlefield arsenal—many attempts are made, of which we are unaware, to create weapons that will kill or mutilate as massively and indiscriminately as nuclear warheads. Consequently, the future of humankind rests with the development of alternative defense technologies. The business of conventional weaponry must be ultimately terminated and retooled in a direction worthy of the human effort. If we are to truly evolve as a species, we must learn to think in terms that will promote our evolution to a higher plane. Philosophically, idealistically, practically—and realistically—we cannot walk the same path to the future lest we find ourselves walking through, and revisiting, our history.

In January 1991, the secretary of defense directed the creation of a U.S. Department of Defense working group on nonlethal weapons technology and its application to the new defense policy.[2] As of this writing, the significance of this historic directive has yet to have an impact on the American people, but the very fact that nonlethality is being considered a viable and critical tool in the projection of force beyond the twenty-first century is remarkable.

This book examines the theory that there is a linkage among the strategic setting, national security strategy, national military strategy, and kinds of weapons and forces that a nation develops to support power projection and achieve those strategic objectives. Further, it portrays a new generation of weaponry designed to shield human life while maintaining the destructive effect of weaponry against war machines. The work is designed to provoke thought and lay a foundation for further exploration.

This book is not a technical description of gadgets, formulas, and computers. Instead, it is the first work to attempt to explain clearly how a nation's forces and weapons are linked to its national security and military strategy. It then uses the analysis to explain why, in a new strategic era, alternative weapons and technologies of a nonlethal class might better serve our national interests. It is also the first book to provide readers with the necessary vision to make decisions regarding the employment of these nonlethal weapons and technologies. In the final chapters, the book takes nonlethal technologies and integrates them into specific scenarios characteristic of current world events, thus illustrating in hypothetical terms how individual technologies might be employed at the tactical and strategic levels.

OBJECTIVE

In response to radical changes, internationally and domestically, that have redefined U.S. national security strategy, U.S. military strategy must include the use of nonlethal weapons technology. Further, nonlethal weapons and technologies must be defined and categorized in order to establish a common view as to their potential application at the tactical and strategic levels. Those definitions follow.

National Security Strategy and Military Strategy Defined

National security strategy and national military strategy are spoken of extensively in the following chapters. To grasp fully the message of volume requires understanding the nature of national security strategy and its link to national military strategy. Clausewitz's famous dictum, "War is simply a continuation of political intercourse, with the addition of other means," is a simple, but provocative, explanation of my aim.[3] In rudimentary terms, strategy is a plan of action that organizes efforts to achieve objectives. Today it is more accurate to describe strategy as a complex decision-making process that ties together a nation's interests, security objectives, and the ways and means employed to achieve those objectives—primarily, weapons and forces.

National interests are simply national needs and wants. These are divided into three categories: vital, critical, and peripheral. The distinction between vital and critical interests is blurred. Former President Richard Nixon clarified it in *Seize the Moment*, where he defined a vital interest as one whose loss "in and of itself directly endangers the security of the United States." Further, Nixon defined a critical interest as one that, if lost to a foreign power, "creates a direct threat to a vital interest." Finally, he described a peripheral interest as one that, if lost, would have a "distant impact on U.S. vital and critical interests."[4]

In addition, vital and critical interests are usually those for which a nation is willing to take risks. Peripheral interests are those that a nation would like to promote, protect, or attain. Human rights may be an example; however, few nations, except in extreme circumstances, are willing to place the nation at risk to secure peripheral interests for someone other than themselves.[5]

National interests can be tangible or intangible. Securing a nation's borders is an example of the former, while the promotion of democracy is an example of the latter. Every nation has core or fundamental interests addressing security, stability, and prosperity. For the United States, security (defined as the protection of the citizens and of key values and ways of life) is a fundamental interest.[6] Regardless of a nation's definition of its core interests, specific interests are derived from generic ones. From these specific interests, a nation defines its national objectives.

The identification of a nation's security objectives is critical to the strategy process. For the United States, national security objectives have revolved around, and are translated into, a strategy of containment and deterrence. The main emphasis of this strategy from 1945 to 1989 was the containment of Soviet-sponsored communism and the deterrence of war; specifically, nuclear and global war.[7]

National security objectives must be well defined, consistent, and supported by some consensus of the elements of national power. Those elements of power include the geographic, political, economic, and military, as well as the national will. If security objectives do not consider the elements of power listed here, the result is usually a failed political, economic, or defense policy.[8] Two Cold War examples are the Korean and Vietnam Wars. Both conflicts pursued objectives under the direction of national leadership, ignored major elements of national power, and failed to secure their objectives.[9]

The elements of power used to secure national objectives may vary from political to economic, but the last recourse is military power. The military element of power is the means through which a nation may focus its national power most directly and most quickly against another nation. Military force is the ultimate statement of national will in international relations and is interrelated with the political element of power.[10]

Military power can be defined as the creation of weapons and forces that are designed to project power and achieve national objectives. Military power, and its use to guarantee the security of a nation, comprise a dominant theme in the study of international relations and national security. Understanding the requirement for military power, its effect on international relations, and the development of national strategy begins with an understanding of the Hobbesian model. In 1651, Thomas Hobbes published *Leviathan*, a classic in which he presented a theoretical foundation of international relations. "Nature," Hobbes said, "has made men so equal" that, while one may be "stronger in body or quicker of mind than another . . . the weakest has strength enough to kill the strongest," either by surprising the victim or by joining with others.[11] The Hobbesian state of nature is a state of constant war, with "every man against every man," and consisting, not only of battle, but also of the psychological instability that results from knowing that some person or group might attack at any moment.[12] Although Hobbes did not extend his description to include the nature of the international community, others have done so.

In the absence of a common power to keep each member of the international community in check, each nation must either be strong enough to ensure its own survival against likely aggressors or form an alliance with those that can.[13] John Spanier's international politics text, *Games Nations Play*, described the international community as a system in which

> Each state . . . is the guardian of its own security. . . . Each regards other states as potential enemies that may threaten fundamental interests. Consequently, states generally feel insecure and regard one another with a good deal of apprehension and distrust. All become very concerned about their strengths, or power. In order to prevent an attack, a state must be as powerful as potential aggressors.[14]

Some might argue that Spanier is extreme in his characterization.[15] However, influential policy makers and strategic planners have used, and continue to use, the Hobbesian model.[16] For example, the model underscores the secretary of defense's current (1991) annual report to the president and Congress and President George Bush's current national security strategy, which are used frequently in this analysis.[17]

Nonlethal Defined

German bombing raids against Great Britain during World War II were conducted mostly at night. The bomber pilots used a radio beam emitted from various ground transmitters to guide them to and from the target areas. These beams provided only azimuth information to the German aircrews.[18] Eventually, British scientists, experimenting with countertechnologies, developed a method of "bending" these beams. By altering the information received in the cockpit by the German aircrews, if only for a brief moment, the British were able to induce a serious navigational error in the bombers. The error was significant and, given the fuel constraints of the era, the bomber crews were forced to return to base without striking their intended targets.[19]

The reader must understand that I define a nonlethal technology as one having the potential for development into weaponry that can disable or destroy an enemy's capability to continue the warring effort without causing significant injury, excessive destruction of personal property, or widespread environmental damage.[20] In essence, it can destroy the enemy's ability to wage war through the destruction of the war machine, thereby ultimately forcing an alternative solution to traditional means of conflict resolution. The U.S. Training and Doctrine Command (TRADOC PAM 525–5, 1 August 1991) further defined nonlethal technologies as those, "typically acting through discriminative vectors that prevent normal operation of equipment and personnel."[21] These weapons can be delivered via numerous autonomous mechanisms, as the consequences of indiscriminate targeting are minimal. The weapons systems are not designed to kinetically destroy people or materials. Examples include low-energy lasers to temporarily blind people, infrasound waves to temporarily disorient and incapacitate them, and chemical agents to change the molecular structure of base metals or alloys on critical aircraft, ships, trucks, or armored vehicles.[22] My argument with this definition is that it leaves the door open for the development of technologies designed specifically to maim the opponent rather than the war machine. Devices permanently blinding or destroying a person's mental capacity are, by definition and class, nonlethal weaponry; however, it is my belief that they should be excluded from the potential nonlethal aresenal for moral reasons, if nothing else. We are exploring more advanced and humane methods of resolving conflicts in the new millennium; consequently, we should not be entertaining technologies designed to cripple and mutilate. To do so would be to repeat history.

There is no universally accepted term for a nonlethal capability, which may be the greatest deterrent to the acceptance and employment of these technologies. For example, the U.S. Department of Defense Advanced Research Projects Agency calls the capability, "mission kill." The U.S. Department of the Army calls these forms of technologies, "nonlethal," and the Department of the Navy has adopted the phrase, "soft kill." However, all these labels refer to the same concept and capability. In this work, the terms *nonlethality* and *nonlethal weapons and technologies* will be used.[23]

This work takes the first steps toward developing a paradigm around which to organize the available information in this area; it proposes a methodology to study it and a vision for its use. Originally, the intent was to devote the greater portion of this study to nonlethal concepts and hypothetical applications of nonlethal technologies. However, it proved arduous to transform a genuine fascination with a concept into a document that would be viewed by readers as something other than an embrace of folly. Therefore, the synthesis requires analysis, which is provided in Chapter 3 ("Historical Model of the Theory: Imperial Japan, 1900–1941"), Chapter 4 ("Historical Model of the Theory: The United States in the Cold War Era, 1945–1989"), and Chapter 5 ("An Era of Transition: The United States in the Post–Cold War, 1989–1995").

The approach adopted is a historical application of a model using a common framework for analysis. First, the study demonstrates the utility of the model with Imperial Japan, showing the linkage among the strategic setting, national security strategy, national military strategy, and the weapons and forces Japan developed to project power in support of the strategies. Then, as a separate application of the model, we will examine Cold War America, followed by post–Cold War America, scrutinizing each era against the model used for Japan.

This framework is founded on the conclusion that the strategic setting, national security strategy, and national military strategy are inextricably linked to the evolution of weapons and forces adopted by nations. By establishing this link, the work demonstrates that the current vision of new world order demands an innovative new weapons concept. For this book, that concept is nonlethality.

Defense policy, Samuel Huntington wrote, is Janus-like, existing in the world of both international and domestic politics.[24] At the same time, it involves a threat frequently generated in the international politico-economic environment, which requires a domestic response or reaction. For example, Glenn H. Snyder has added a further dimension by arguing that national strategy and security policy involve two concepts, deterrence and defense:

> Essentially, deterrence means discouraging the enemy from taking military action by posing for him a prospect of cost and risk which outweighs his prospective gain. Defense means reducing our own prospective costs and risks in the event that deterrence fails. Deterrence works on the enemy's intentions; the deterrent value of military forces is their effect on reducing the likelihood of enemy military moves.
>
> Defense reduces the enemy's capability to damage or deprive us; the defense value of military forces is their effect in mitigating the adverse consequences for us of possible enemy moves, whether such consequences are counted as losses of territory or war damage. . . .
>
> Perhaps the crucial difference between deterrence and defense is that deterrence is primarily a peacetime objective while defense is a wartime value. Deterrence and defense value are directly employed in different time periods.[25]

Given these concerns regarding national security policy, this study designed a four-part framework to analyze the linkage between defense policy and the evolution of conventional weaponry. The four parts are: the international and domestic environment, often referred to in this work as the strategic setting; the evolution of national security objectives; the evolution of national military strategy; and the nature and evolution of weapons technologies to project the national defense policy.

A FRAMEWORK FOR ANALYSIS

The defense policy of any nation-state is tied closely to its position within the international environment of interacting states.[26] This international environment, as well as the domestic perception of it, can be understood as both the source of opportunities that a state may wish to pursue and the source of threats to its security. Given this explanation, it is certain that if we are to study the defense policy of the United States, the study should begin with an assessment of the U.S. position in the international system and its own domestic setting.

The second part of the organizational framework conforms closely to a rational actor model of decision making on national security strategy. That is, decision makers acting for a state establish doctrine on national strategy and military force employment based in part on national objectives, rational capabilities, and constraints. A counterargument to this claim is that national strategy, force structure, military doctrine, and other elements of national security policy do not necessarily conform to this pattern. In short, is strategy actually linked to the evolution of weapons technologies, or does the latter determine the former? To explore this argument is beyond the scope of this book; however, the controversy is identified for readers who may wish to explore it.

In the third part of the framework, the book addresses the issue of national military strategy and its linkage to national security strategy. In a basic sense, why does the military strategy of a nation conform to, and so closely support, the national security strategy? What is the relative importance of the international and domestic settings, to the national security strategy in determining military strategy? This book will not attempt a detailed analysis of all the elements of security strategy and military strategy, but sufficient parallels are drawn to make clear the relationship between the two.

Finally, this work addresses the evolution of weapons and forces designed to support a changing national security strategy. This portion of the framework illustrates how a nation's choice of weapons and forces are linked to its military strategy. The illustration of weapons and forces is generic by design but detailed enough to clearly reveal how they evolve to meet the needs of national security policy. The framework in its entirety is provided here so that readers can understand the detail applied in the analysis.[27]

The Analytical Framework

A. The International and Domestic Setting—Strategic Setting
   1. Power and status
   2. Relative capabilities
   3. Threats:
      a. International and external threats
      b. Nature of the threat (military, political, economic, and cultural)
      c. Perceptions of the threat by decision makers
   4. Linkages and interdependencies

B. National Security Strategy
   1. National security objectives
      a. Changes in national security objectives
      b. National will
   2. Domestic determinants of national strategy
      a. Political system
      b. Economic system
      c. Geography
      d. Public opinion (national will)
      e. Technology
   3. Capabilities for accomplishing national security objectives
      a. Military capabilities
      b. Economic capabilities
      c. Technological capabilities
      d. Political capabilities

C. National military strategy
   1. National military objectives

2. Manpower
3. Industrial capacity
4. Economic and budgetary limitations
5. The relationship among the objectives, strategy, and means (1–4)

D. Weapons and Forces (the Means)
1. Military hardware and force structure
2. The use of force
   a. Technological developments and the use of force
   b. Objectives sought through the use of force

A NEW AND EMERGING CONCEPT

Combat usually attempts to render enemy forces incapable of hostile action. Until recently, destructive mechanical means have been the weapons of war, but now, a technological revolution has begun that will rival the discovery of atomic weaponry. This revolution is based on the emergence of an alternative to antipersonnel weapons technologies and systems, and focuses instead on antimatériel technologies that are termed nonlethal.

The concept of weapons that are antimatériel, as opposed to antipersonnel, will significantly alter the manner in which the nation's military will fight. The minute specifics of this alteration are not addressed in this book. Some of the more critical issues relating to this alteration are discussed in Chapter 7 ("Epilogue"), but the reader should understand that these are minimal claims, which are not designed to offer answers but rather to provoke further study and thought. Furthermore, I would be naive to claim that nonlethal (antimatériel) will *replace* lethal (antipersonnel technology), at least in the near term.[28] At best, we can hope these technologies will be accepted as offering tremendous potential to global conflict resolution, perhaps prompting the acquisition of a handful into the existing convention arsenal. In the long term, as Senator Alan Cranston identified the issue in the first annual in the Mikhail Gorbachev State of the World Forum in 1995, we may hope for the establishment of a global peacekeeping force armed with nonlethal weaponry. This is the dream—this is the goal.

Although dynamic, the evolution of this concept has been slow, and only a small number of military personnel are aware of it. Few have considered its tactical implications, and even fewer have entertained it as a potential global strategy for the twenty-first century. Nonetheless, Janet Morris, of the U.S. Global Security Council, called the theory of nonlethality, "a revolutionary concept" that will allow the United States to lead the international community into a new world order. Further, it will move those who grasp it away from waging war and toward a more positive national strategy of peacekeeping. In addition, this "new way of war'' will continue to enhance our diplomatic efforts and improve our ability to project U.S. power, when necessary, worldwide.[29]

The United States currently has the technological capability to pursue its national security objectives, abroad and at home. For example, hydrogen embrittlement, ultraviolet ''elastic'' light distancing and ranging (UVE-LIDAR), engineered computer viruses, bioengineered plant pathogens, infrasound, calmative agents, super-polymers, and super-caustics are just a sampling of the technologies currently in development at various labs throughout the nation. Some technologies in pursuit of defense objectives enjoy government funding, while others wait for funding to be fully employed.[30]

While the United States and other global powers strengthen their powerful deterrence capability with high-technology, ''smart'' weapons of the lethal variety, they now have an obligation to explore the role of nonlethal technologies in support of, and, in some cases in lieu of, lethal technologies. These bold claims are examined in detail in Chapters 3, 4, and 5. For now, consider that the United States has an opportunity to adopt this new policy, strategy, and operational capability before another nation preempts it and gains the invaluable media and diplomatic advantage of being the first to delineate this life-conserving strategic philosophy.

LIMITATIONS

Documentation for this book is limited to unclassified sources; however, the information necessary to construct the hypothetical scenarios used to illustrate potential uses is, in many cases, classified. I am willing to accept responsibility for the use of this information, having used the following criteria for that decision: does what I am saying in this work in any way endanger the national security of the United States or the lives of those affiliated with these technologies?

I struggled with these questions; however, after years of working in the intelligence and special operation communities of the U.S. Army, I have reached the conclusion that there are certain things the world has a right to know. Those rights are cut short when national interests or lives are at stake—but that is not the case here. I am not giving away launch codes or target information; nor am I providing the names and addresses and covers of U.S. operatives. I am not detailing the specific technical data related to any of the technologies; instead, the information provided here is generic. The only specifics relate to the possible employment capabilities of the technologies, and in some cases, even those have been altered. Further, each of the technologies used in the scenarios has previously been released or mentioned in open-source literature;

only the vision of the application that has been missing—which I provide.

This book addresses hypothetical applications of nonlethal technologies and their proposed effect in a continued deterrent role in support of national security and military objectives. A more detailed comparison and analysis of lethal and nonlethal technology is a germane issue, but that is beyond the scope and intent of this book.

In the realm of nonlethal technologies, sufficient information is not available on hostile countermeasures. To discuss the nonlethal technologies in the Chapter 6 scenarios, I begin with a basis of fact, beyond which I can only speculate. Therefore, hostile countermeasures and their effect are assumed or omitted depending on the technology. The evolution of this concept has been so rapid that in many cases, the existing technologies pioneered in the United States have not explored countermeasures. For this reason, discussion is limited.

In Chapter 6 ("The Vision"), the evolution of nonlethal technology, its potential application and impact on emerging U.S. national security strategy and military strategy are discussed in scenario form. Of the fourteen existing nonlethal technologies, which include the biocontrol of narcotics, hydrogen embrittlement, super-caustics, super-polymers, combustion modifiers, high-powered microwaves, directed-energy (laser) technologies, antitraction devices, nonlethal ballistics (the shroud round and rubber bullets), infrasound, computer viruses, and others are characterized and discussed as they might relate to: operation Desert Storm, Bosnia, the Middle East, Latin America, the counterdrug effort, urban encounters, and weapons proliferation scenarios. A more specific description of these technologies follows. The reader must understand that for security reasons I have either omitted by design some of the characteristics and specifics of these potential weapons or have been given only partial information about these technologies from the engineers, scientists, and acquisition personnel with whom I spoke.

I certainly am willing to provide more than anyone else would: that is, if I feel it warranted to give you classified information in order to make you a believer, I will do so. I will not give classified information that would be detrimental to the overall goal of this book or of individual technologies; you must trust my judgment. Philosophically, I am at an impasse with the defense community. I believe there are secrets worth keeping, however, I also believe there are secrets that should be revealed because they potentially could change the world for the better. There is another drawback to telling this story about these technologies. Much of what we are talking about here comes from the protective cover of the intelligence community. As a friend of mine, former Army Master Sergeant Lyn Buchannon, is fond of saying:

When dealing with classified issues arising from the intelligence community, one must recognize that every truth is part lie and every lie is based on some truth. If you try hard you might be able to separate the two. If you're lucky and work hard, you will find some of the truth. . . . If you're lucky and work really hard, you might find the whole truth as somebody wants you to know it! If you are phenomenally lucky and really work your tail off, you might even go on to find the real truth; but, no outsider, and in fact very few insiders, ever learn the whole real truth.

With that caveat, here are the technologies; in Chapter 6, you will find their applications.

The Weapons and Technologies

- *Hydrogen Embrittlement:* Incorporates the use of microencapsulated liquid hydrogen, or hydrogen in any form. The substance is applied to lethal weapons or generic war machines to "embrittle" the surface of the target, rendering it ineffective as a weapon. When fired or used, the lethal weapon or machine would explode or crack from the application of the hydrogen.
- *Calmative Agents:* Gases, liquids, pills and optic devices designed to induce a relaxed or slumber state. These technologies can be employed en masse or on select individuals.
- *Antitraction:* These devices and technology can be aerosol applied, poured, or painted on any surface, rendering it slippery or boggy. Some of these technologies will lubricate the surface, preventing traction of people and machines; others will soak into the surface on which they are applied, making a chemical mud that prevents movement of personnel or equipment.
- *Super-Caustics:* Various substances applied or introduced to lethal weapons or war machines that will destroy the metals or composites used in their manufacture. The effect is corrosion, embrittlement, evaporation, and the like.
- *Super-Polymers:* Various substances applied by aerosol, liquid, or powder form that will (when conditions dictate) form an impenetrable surface over the target. The effect is like applying a permanent, quick-drying glue over an entire target, rendering all its moving components ineffective.
- *Combustion Modifiers:* Various gases, foams, or liquids that are designed to choke internal combustion engines. The effect is the elimination of any running engine on the battlefield. There are also additives that will corrode the internal components of any engine that aspirates the substance, rendering the prime movers of any warring faction "dead in the field."

Nonlethal Ballistics

- *Shroud Rounds:* Tank-, artillery-, or aircraft-delivered munitions that explode milliseconds before impacting their intended target, forming a polymer sheet that is reinforced with a fine aluminum wire mesh. The effect is a weapon that envelopes its target much like an octopus, wrapping it in the polymer sheet and thus shorting out its communications and optics and choking its engines.

- *Rubber Bullets:* Bullets designed to sting but not kill. Employed in various sizes and shapes, they can stop individuals and war machines through the infliction of temporary pain or the destruction of various mechanical components, respectively.
- *Infrasound:* Employs sound waves in a coherent or concentrated form against various targets. The effect is the shattering of metal or composite materials on war machines or building materials.
- *High-Powered Microwaves (HPM):* Designed to disrupt brainwave patterns, communications, or any electronic component of a machine. The effect is confusion, stupor, or coma in people or animals and the disruption, scrambling, or jamming of electronics.
- *Project Flicker:* A candle or light source that will operate at a predetermined frequency and pulse rate and will induce vomiting, sickness, nausea, slumber, in its target.
- *Directed-Energy (DE) Weapons:* Any coherent or concentrated energy source (e.g., lasers). The effect is the burning, cracking, distortion, or impairment of conventional or unconventional war machines. In people these technologies can be preset to invoke stammer, confusion, or coma.
- *Microbic Technology:* Technologies that essentially create designer weapons using living microbes as the destructive force. These weapons can be delivered and introduced to the battlefield via a number of vectors (potable liquids, fuels, solids, and aerosols). Each specific suite of these organisms is designed to attack a specific target. For example, a particular microbe can exist for long periods of time immersed in liquid aircraft fuel. However, when the fuel is ingested by the aircraft, the microbes immediately begin consuming the plastic materials used in the aircraft's fuel filters. Within a relatively short period of time, the aircraft is grounded—as is the rest of the squadron, if they used the same contaminated fuel.
- *Computer Virus Technology:* Various destructive or directive viruses that can be actively or passively introduced into the target. They can be cross-employed by HPM weapons and DE weapons. The effect is the impairment, destruction, or redirect of the intended target, electronics, computers, or other high-end technologies.

In the future, specific nonlethal technologies should receive more detailed analytical study in order to better access the impact of their application on defense policy objectives. The limitations on the definition of what is nonlethal and on the availability of useful studies and reports are covered in the following chapter.

NOTES

1. Sun Tzu, *The Art of War*, trans. Samuel B. Griffith (New York: Oxford University Press, 1963), 77–79.
2. This information was extracted from one of the promulgating documents

for nonlethal concepts. The referenced memorandum is signed by the undersecretary of defense, Paul Wolfowitz, and dated 30 March 1991. The subject heading is: "Do We Need a Nonlethal Defense Initiative?" Marginalia in the pen of Secretary of Defense Dick Cheney approved the recommendation to convene a strategy group and explore the concept of nonlethality.

3. Carl von Clausewitz, *On War* (1832; reprint, ed. and trans. Michael Howard and Peter Paret, Princeton, NJ: Princeton University Press, 1976), 1: 7.

4. Richard Nixon, *Seize the Moment* (New York: Simon and Schuster, 1992), 36–37.

5. U.S. Directorate of Joint and Combined Environments, "Military Power and National Objectives" (C500, Lesson 1, 1991–1992), 8–10. Lesson plan used in the curriculum of the U.S. Army Command and General Staff College, Fort Leavenworth, KS.

6. Ibid., 9.

7. For a broader understanding of the post–World War II controversy concerning the Soviet threat and the proper U.S. response, see the following articles: George Kennan (as "X"), "The Sources of Soviet Conduct," *Foreign Affairs* (July 1947): 566–582; John Lewis Gaddis and Paul Nitze, "NSC 68 and the Soviet Threat Reconsidered," *International Security* (Spring 1980): 164–176; Editors of *Foreign Policy*, "Interview with George Kennan: 'X' plus 25," *Foreign Policy* (Summer 1972): 5–21; Charles Gati, "What Containment Meant," *Foreign Policy* (Summer 1972): 22–40; Charles M. Roberts, "How Containment Worked," *Foreign Policy* (Summer 1972): 41–53; George Kennan, "Reflections: Two Views of the Soviet Problem," *New Yorker*, 2 November 1981, 54–62.

8. U.S. Directorate of Joint and Combined Environments, "Military and National Objectives," 8–10.

9. Ibid., 24–25.

10. Ibid.

11. Thomas Hobbes, *Leviathan*, ed. Herbert W. Schneider (Indianapolis, IN: Bobbs-Merrill Educational Publishing, 1958), 104–105.

12. LTC James M. Dubik, USA, "The Foundations of National Strategy: Past and Present" (First Term, AY 90–91), 11–13. Unpublished monograph for the School of Advanced Military Studies, United States Command and General Staff College, Fort Leavenworth, KS.

13. Former Soviet president Mikhail Gorbachev took the lectern at Westminster College in Fulton, Missouri, on 6 May 1992 to declare the Cold War won and promote his vision of a global society. Gorbachev advocated the formation of a "global government" with a "special body" (common power) under the United Nations having the power to enforce political, economic, diplomatic, and military solutions to regional conflicts.

14. John Spanier, *Games Nations Play*, 5th ed. (New York: Praeger Publishers, 1975), 11.

15. Charles Beitz, *Political Theory and International Relations* (Princeton, NJ: Princeton University Press, 1979), 27–50.

16. For a more contemporary use of the Hobbesian model, see John J. Mearsheimer, "Why We Will Soon Miss the Cold War," *Atlantic* (August 1990): 35–50. Mearsheimer wrote, "For untamed anarchy is what Europe knew in the forty-five years of this century before the Cold War, and untamed anarchy—Hobbes's

war of all against all—is a prime cause of armed conflict.'' He went on to analyze the demise of the bipolar world as a return to a state system whose character, which is multipolar and unstable, ''drives states toward war.'' This character, he explains, using several scenarios, is quite Hobbesian based.

17. Dubik, *The Foundations of National Strategy*, 11–13; Secretary of Defense Dick Cheney, *Annual Report to the President and the Congress* (Washington, DC: U.S. Government Printing Office, 1991), 1, 5; see also President George Bush, *National Security Strategy of the United States* (Washington, DC: U.S. Government Printing Office, 1991), 15, 23–50.

18. Herbert H. Dobbs et al., ''Assessment of Mission Kill Concept, Requirements, and Technologies,'' SPC Log No. 90–1987 (Arlington, VA: Defense Advanced Research Projects Agency, September 1990), 6.

19. Ibid.

20. U.S. Army Training and Doctrine Command (TRADOC), *AirLand Operations: A Concept for the Evolution of AirLand Battle for the Strategic Army of the 1990s and Beyond*, TRADOC PAM 525–5 (Washington DC: U.S. Government Printing Office, 1 August 1991), 48.

21. Ibid.

22. Ibid.

23. Dobbs et al., *Assessment of Mission Kill Concept*, 6–8.

24. Samuel Huntington is a defense analyst and author of *The Soldier and the State* (Cambridge: Harvard University Press, 1957). I have used the terms *defense policy* and *national security policy* interchangeably in this work. Argument can be made against this practice; however, to do so best served my analysis. See Samuel Huntington, *The Common Defense* (New York: Columbia University Press, 1961), 268.

25. Glenn H. Snyder, ''Deterrence and Defense: A Theoretical Introduction,'' in *American Defense Policy*, ed. John E. Endicott and Roy W. Stafford, Jr., 4th ed. (Baltimore, MD: Johns Hopkins University Press, 1977), 39–40.

26. Michael Douglas and Paul Viotti, eds., *The Defense Policies of Nations* (Baltimore, MD: Johns Hopkins University Press, 1982), 4–5.

27. The framework provided here is altered slightly for analysis of the United States and Japan. There were several areas where the framework would not apply due to changes in the ideological nature of the two countries.

28. There is little doubt that beyond the twenty-first century, warfare will focus on the elimination of an enemy's ability to wage war. Other means may be required to degrade the enemy's will to fight, but the elimination of a war machine is within our grasp. Lynn Buchannon, president of a Maryland-based company, P.S.I., Inc., is in the business of making predictions on global competitive strategies for Fortune 500 clients. P.S.I., Inc., claims that future weapons will be dynamic; in Buchannon's words, ''These new families of weapons will stretch all definitions of normalcy, similar to, but exceeding, the revolutionary concepts of atomic weapons in the mid-1940s.'' Further, Buchannon said, ''World focus will turn toward more peaceful issues of technology transfer and environmental restoration [instead of] the continued development of weapons of mass or surgical destruction.'' Interview conducted by the author in Buchannon's home in Mechanicsville, Maryland, on 5 January 1990. Notes in author's collection.

29. The U.S. Global Security Council (GSC) is a nonprofit, educational research

organization, incorporated in the state of Virginia. The information quoted in this study is extracted in part from notes taken during a (GSC) briefing to the U.S. Army Training and Doctrine Command (TRADOC) in March 1990, at Fort Monroe, Virginia. Notes in author's collection.

30. In Chapter 6, several nonlethal technologies are explained in as much detail as possible given the limitations of the analysis. However, off-the-shelf nonlethal technologies are nothing new, as the application of some technologies in a nonlethal role may indeed be something unintended yet applicable. Liquid hydrogen embrittlement, simply explained, is the application of liquid hydrogen to a military target using any number of delivery means; remotely piloted vehicles (RPVs) and aerosol application are two examples. The hydrogen freezes the surface materials of the target on contact, making it brittle. In test cases, the target material became brittle enough to shatter when struck with a foreign object or, in the case of a tank main gun, when fired. Reports of such testing are doubtless classified, and I mention it here in an unclassified explanation only to serve as an example.

Chapter 2

The Current Position

This chapter reviews the status, in literature and actions, of developments in nonlethal weapons and technologies. The review focuses on two functional areas: first, nonlethal technology and weapons, and second, global, national, and military strategy. Six categories of investigative material were used: books, government documents, periodicals, unpublished materials, interviews, and completed video projects from sources like the Center for Defense Information in Washington, DC. Not surprisingly, an investigation of the research question revealed a sizable amount of material pertaining to the U.S. security strategy, old and new.

Literature in any form pertaining to nonlethality is difficult to find. Primarily, this is the result of the relative immature state of the concept; also, however, the idea has few supporters. Documents dealing with technologies that could theoretically be molded to nonlethal applications are available. For example, studies in lethal-agent chemical warfare conducted by the U.S. Air Force Armament Lab, at Eglin Air Force Base, Florida, have identified spin-off technologies, such as incapacitating agents, that can be adapted to nonlethal roles. However, these reports and documents are research oriented, highly technical, and parochial. The reports address technical specifications. They do not support hypothetical applications in scenarios, nor do they address the use of the technology in a nonlethal form.

In another example, studies conducted by the U.S. Department of Defense Intelligence Agency's Technical Intelligence Office have explored weapons delivery systems for lethal chemical agents. One delivery mechanism mentioned in the study is microencapsulation, a process that is, in itself, a science. Very basically, it involves the jacketing or covering of different materials; the capsule can, in turn, be engineered (the encapsulation) to release its contents under varying conditions. The trigger can be anything from time to contact with human skin; the process is analogous to a plastic bag containing gasoline being held over the flame of an ordinary match. In this example, the plastic encapsulator may be specifically designed to protect the flammable liquid within, and under these temperatures, the fuel will not ignite even though the exterior of the container exceeds the flash point of the fuel. However, when the bag is held over a common laboratory Bunsen burner, where the temperature of the flame can reach 1600°C (about 2910°F)—hypothetically the trigger temperature of the encapsulator—the material surrounding the fuel will dissolve or disintegrate, thus exposing the fuel to the flame. The results are obvious. Theoretically, the trigger can be tuned to any temperature, giving it a wide range of applications. Therefore, microencapsulation as used to contain certain nonlethal chemical or biological agents can have a nonlethal role, just as it has many other scientific and practical roles. With that understanding, microencapsulation is described (in Chapter 6) as being used in conjunction with certain agents in a nonlethal role.

An exhaustive search for reports, articles, or studies was conducted using the Combined Arms Research Library's public access module, the Defense Technical Information Center's on-line system (DROLS), and Infotrack. Extensive research was conducted on the Internet, where I explored various Department of Defense web sites and countless science and technology research facilities, as well as material by freelance authors, organizations, and groups expressing an interest in global weapons issues. The search was limited to the last ten years and requested information on the weapons and technologies of the nonlethal, antimatériel, soft-kill, and mission-kill categories. Unfortunately, I obtained little or nothing of value under any of these classifications.

Three reports obtained through personal contacts have "nonlethal" in the title and discuss nonlethal weapons; however, these were misleading. The studies are: *Nonlethal and Nondestructive Combat in Cities Overseas*, by the Institute for Defense Analysis, Science and Technology Division (IDASTD); *Report on the Attorney General's Conference on Less Than Lethal Weapons*, by the U.S. Department of Justice (USDJ); and *Assessment of Mission Kill Concept, Requirements, and Technologies*, prepared by the Defense Advanced Research Projects Agency (DARPA).

The reports on studies conducted by the Department of Justice and the Defense Analysis Science and Technology Division dealt with weapons and technologies available to law enforcement officials. For example, these studies were limited to riot batons, personnel nets, handcuffs, stun guns, and so forth, making them insignificant to this study.

The report from DARPA looked at nonlethal technologies in a military role, for example: antimatériel technologies, mechanical devices, chemical agents, electronic devices, and optical devices. The report was useful; however, it was of minimal value to this study, primarily because of its limited scope and abbreviated format. The authors stated that "the findings and recommendations of the committee are directed at the management of current research."[1] Therefore, they did not focus on potential applications, nor did they address threat countertechnologies. The report is cited for definition, but its value to this study is finite.

UNPUBLISHED MATERIAL

I should make clear my realization that the work of most of the proponents of nonlethal systems is related to weapons systems and the operation of the physical laws of nature in the context of sophisticated military technology. To the extent that their work is rigorously scientific, they have every right to be impatient with the dabbling of unqualified nonscientists like me in their technical field of expertise, especially, when I take their life's work and base its significance, in part, on potential future applications to national security strategy and military strategy.

Thus, what appears in this book should not be construed as being critical, either by omission of valid nonlethal technologies or exclusions bodies of reference that may research a discipline of the utmost significance to the security of the United States. In reading, it is clear that the scientists who develop these critical technical and research materials are understandably impressed by the magnitude of the changes over the past decade, as represented by modern developments in nonlethal weapons and related technology. Many of these people—who do not represent all of the American Defense Research Community but probably constitute a significant portion—are convinced that the adaptation of nonlethal forms of weaponry is essential to our survival, not in modern war, but as a world.[2]

Much of the unpublished material written by the proponents of nonlethal weapons and the scientists involved in nonlethal research and development expressed an interest in simulation.[3] A majority of the authors felt very strongly about simulation or modeling as a method of validating the potential of any given discipline. To them the only possible way to plan scientifically for the future is to take the results of testing and evaluation of nonlethal systems and technologies and around it build theoretical mathematical models based on a combination of game theory and educated forecasting of future trends. For instance, models must be able to project forward the effectiveness of the new nonlethal weapons on the basis of comparison with existing information about the past performance of conventional weapons. It is the contention of these scientists that this is the only valid method of determining which technologies are best suited for adaptation into the defense structure—and to obtain additional research and development funding.

On a recent application of simulation, Richard Vitali, a Fort Leavenworth simulations specialist working for the Training and Doctrine Commands, Combined Arms Threats Directorate, told an *Army Times* writer (during the Technology Base Seminar War game held at Fort Leavenworth in October 1990), "The players focused on a Middle Eastern scenario that provided opportunities for using nonlethal [technology]." Vitali mentioned the concept of "incapacitating oil wells so they couldn't be used against us [the coalition forces], but could eventually be restarted again."[4] After the Iraqi forces sabotaged hundreds of oil wells in Kuwait, "the war-game scenario, with its theoretical systems to prevent such destruction, seemed prophetic."[5]

PERIODICALS

Sources of information in this category were almost completely devoted to published articles of the scientific community found in such publications as *Aerospace and Defense Science, Armed Forces Journal—International*, and *Defense Electronics*. These articles did little to support the focus of this book; however, they did contribute to my overall awareness of the revolutionary importance of these new technologies.

The focus of most defense authors, whether or not on technology, continues to lean heavily on issues involved in the creation of multidimensional weapons platforms designed to "meet the ever increasing needs to quickly put into place," anywhere in the world, "forces sufficient to claim credibly the means for sudden massive and decisive action with minimum risk to engaged forces."[6] With these factors in mind, and considering the realities of a constrained budget, the search continues for nonlethal technologies that will support the ability to respond to a broad spectrum of conflict. These technologies must also retain the ability to support the conduct of conventional war against a traditional enemy.

As technology emerges, its potential nonlethal applications becomes an ever-increasing objective of the developers. For example, the recent design of a simple microchip has a possible nonlethal (defensive form) application. Terrorists getting their hands on tube-launched, optically sighted, wire-guided (TOW) "Stinger" missiles of the future may well find those otherwise lethal weapons useless, thanks to a proposed fire-enabling security chip that "is getting serious attention [from] Capitol Hill."[7] The chip is designed to require an enabling code before it will permit firing of the weapon. The chip can also be designed to disable the weapon from a distance, and best of all, the chip is presumably beyond the capability of Third World countries to duplicate. This is further evidence that the scientific community realizes the need for focused efforts in the realm of nonlethal technology.

At first, enabling chips may seem not a nonlethal technology, but rather a security measure to prevent the system's unauthorized detonation, not unlike those found on nuclear devices throughout the world. However, if this technology is introduced in a "strategic deception" role, that is, if it is sold to unwitting developers of weapons systems, then, based on the current definition of nonlethal technology, enabling chips take on a nonlethal role with unlimited possibilities.[8]

BOOKS

Books were of little value as a source of usable information on the subject of nonlethality per se; many are still in the process of being written, while others circumvent the issues by politicizing the effort. Much of what is written either criticizes the potential of nonlethal weapons or wraps itself in politics and employment strategy without ever rendering statements, opinions, or facts about the direction of the defense industry and the immediate need for philosophical reform. However, books were invaluable to my study of concepts, national security strategies, and military strategies and their relationships to weapons and forces. I could not have established a global historical perspective through research only using periodicals, unpublished works, and executive memoranda. To develop an understanding of alternative perspectives on the causes of war and conditions of lasting peace, I searched books on a wide spectrum of related topics.

Issues on national defense policy and unilateral disarmament were researched in detail in the works cited. Other books used focused on national posture and how we prepared to seek alternatives to war. This volume incorporates, in large part, information derived from investigations of other authors on the future of humankind. Martin Van Creveld's book, *The Transformation of War*, was a valuable guide in analyzing warfare and strategic issues. Van Creveld's work "aims at providing a new non-Clausewitzian framework for thinking about war, while at the same time trying to look into its future."[9]

Two other books used extensively in forming a background for this work were Robert L. O'Connell's *Of Arms and Men* and Antonia H. Chayes and Paul Doty's edited volume, *Defending Deterrence*.[10] Both analyze historical cases of extended deterrence, using them to illustrate those that face the United States today. The authors formulate and test hypotheses on linkages between political and military conditions under which deterrence is likely to succeed or fail in the new world order. Other critical authors to the background are Alvin Toffler and Heidi Toffler, Robert Kaplan, Paul Kennedy, and James Dunnigan, to whom I am grateful for their dedication, knowledge and efforts to educate those willing to explore.

The study uses this information as a basis from which to infer that what has already occurred politically, strategically, and militarily can be used to determine what will occur in these areas in the future. The analysis of how the balance of military capabilities (lethal or nonlethal), alternative strategies of bargaining, past behavior, and issues at stake relate to nation credibility in a deterrent role are used extensively in Chapter 5.

A 1992 book by former deputy director of the U.S. State Department's Policy and Planning Staff, Francis Fukuyama, entitled *The End of History and the Last Man*, provided enormous insight to the linkage of a nation's political evolution, the environment for war, and the process by which humanity and time cope with the dilemma of world affairs. Fukuyama's vision and understanding of the era inspired my fascination with seeking alternatives to the way in which wars will be fought in the future.

Fukuyama's book is not technically oriented. However, it provides evidence in support of the suggestion that there are two powerful influences at work in human history, "the logic of modern science" and "the struggle for recognition."[11] The first, which is linked to economics, outlines humanity's continuous struggle to resolve political issues; the second, "the struggle for recognition," parallels the concepts of Georg Hegel, in describing what Fukuyama calls the "motor of history."[12] Thus, if one is to understand the relationship between society, transnational politics, and war, this book is essential. Fukuyama's book lacks empirical data; however, it was priceless in assisting my efforts to formulate and justify futuristic concepts.

Two other books that I used are the sixth edition of *American Defense Policy*, edited by Schuyler Foerester and Edward Wright, and Marc Trachtenberg's *History and Strategy*.[13] Both proved invaluable in the establishment of evidence supporting the theory that national security strategy is linked to military strategy and that weapons technology parallels the evolution of these strategies.

GOVERNMENT DOCUMENTS

One of the fundamental hurdles to a greater application of nonlethal systems is the lack of formal demand (service requirements) for them. This is readily apparent, as almost all the available government documentation mentions nonlethality only superficially. Without top-level interest, to the degree that nonlethality is discussed doctrinally and is developed into an encompassing program or initiative, nonlethal weapons will likely expand in service interest only glacially.

Training and Doctrine Command (TRADOC) Pamphlet 525–5, *Airland*

Operations, states, "A major opportunity exists in nonlethal technologies."[14] Aside from what is specified, this document is the clear leader in the establishment of a framework in which to apply nonlethal technologies. The implications of nonlethal systems can be used as a model along with the doctrinal framework outlined. It becomes clear from *Airland Operations*, that significant and fundamental changes will likely be required in formal service doctrine to incorporate the concepts of the nonlethal application of force. This will include a consideration of hardware as it becomes available and, more important, the tactical and strategic implications of this new dimension of conflict. The important issue here is to press on with technology opportunities, as discussed in Chapter 6 ("The Vision"); the doctrinal issues will follow. At least one government document entertains the idea.

NOTES

1. Herbert H. Dobbs et al., "Assessment of Mission Kill Concept, Requirements, and Technologies," SPC Log No. 90–1987 (Arlington, VA: Defense Advanced Research Projects Agency, September 1990), 3.

2. The term *American defense research community* pertains to agencies, whether government or private sector, that are involved in the research, development, manufacturing, and sales of defense-related items, including hardware, software, literature, and concepts. It is generally accepted that this community consists in part, of some of the following: the Defense Science Board/OUSA(A); the army's Defense Electronics Command; and any contractor it employs for research or procurement purposes, such as, the Magnavox Government and Industrial Electronics Company, COMSAT, Los Alamos National Laboratory, Defense Logistics Agency, Defense Technical Information Center, Air Force Wright Aeronautical Laboratory, and U.S. Naval Air and Surface Developments Center. To obtain more listings, readers can obtain source documents in defense libraries such as the Combined Arms Research Library (CARL), in Fort Leavenworth, Kansas. The opinion is mine and is based on conversations held and notes taken during discussions on the subject of nonlethality. No sources can be cited.

3. Simulation in this example refers to a process of making quantified judgments based on an analytical model. Simulation in other situations can occur via computer visuals as a feedback of random numerical computer analysis. Formulas and symbols are used to direct data into the analysis. Some of the questions inherent to simulation involve data reliability, interpretation of data, and model fabrication. For more on simulation and gaming, I recommend T. N. Depuy, *Numbers, Predictions and War* (Fairfax, VA: Hero Books, 1985). In my example, simulation refers to the use of computer models to interpret and predict the outcome of any given battlefield situation. The inherent problems of simulation mentioned here, remain constant regardless of the technique.

4. Vitali did not specify just how these oil wells would be incapacitated. However, one would assume that some form of temporary—perhaps subsurface—blocking device would be applied to the wellheads to prevent the use of "flame" to disrupt the environment or cause obscuration via smoke.

5. S. D. Naylor, "Shoot to Disable," *Army Times*, 7 October 1991, 26.

6. Naval Warfare, "U.S. Navy Technology in the '90s," *Aerospace and Defense Science* (Spring 1991): 32–33.

7. M. Struck, "Weapons Security Chip to Get More Hill Attention," *Defense Electronics* 23(10) (October 1991): 5.

8. Strategic deception is a complex and convoluted process of deceiving one's enemy. This practice is employed by the United States to varying degrees. In most cases the concept is classified, if for no other reason than to protect the identity of those involved in deceptions of this nature. I doubt the rest of the world would be shocked to find the United States implicated in these practices. Deception in this realm involves industrial, economic, technological, political, and even human rights targets. It is a distasteful nasty practice, but it opens unlimited conduits for the introduction of nonlethal technologies. The application of nonlethal technologies into an unwitting environment violates the concept of deterrence because the target does not know of the existence of the technology.

The objective of nonlethal technologies is to eliminate an enemy's ability to wage war through the destruction of the war machine. These technologies are sometimes referred to as antimatériel or soft-kill. This is obviously a generic definition of a very multifaceted concept, but it will suffice for one's understanding of using existing technologies, designed for other applications and redirecting those technologies toward nonlethal roles.

9. M. Van Creveld, *The Transformation of War* (New York: Free Press, 1991), 125.

10. R. L. O'Connell, *Of Arms and Men: A History of War, Weapons, and Aggression* (New York: Oxford University Press, 1989); and Antonia H. Chayes and Paul Doty, eds., *Defending Deterrence: Managing the ABM Treaty Regime into the 21st Century* (New York: Pergamon-Brassey's International Defense Publishers, 1989).

11. F. Fukuyama, *The End of History and The Last Man* (New York: Free Press, 1992), 71–81, 143–208.

12. George Wilhelm Friedrich Hegel, an eighteenth-century philosopher, was noted for his efforts to write a "Universal History" and other works on collective human consciousness and the "dialectic" process of change.

13. S. Foerester and E. N. Wright, eds., *American Defense Policy*, 6th ed. (Baltimore, MD: Johns Hopkins University Press, 1990); and Marc Trachtenberg, *History and Strategy* (Princeton, NJ: Princeton University Press, 1991).

14. U.S. Army Training and Doctrine Command, *AirLand Operations: A Concept for the Evolution of AirLand Battle for the Strategic Army of the 1990s and Beyond*, TRADOC PAM 525–5 (Washington, DC: U.S. Government Printing Office, 1 August 1991).

Chapter 3

Historical Model of the Theory: Imperial Japan, 1900–1941

This chapter will illustrate the linkage between Imperial Japan's strategic situation, national security strategy, national military strategy, and the kinds of weapons and forces developed to project power and achieve its strategic objectives.

JAPAN'S STRATEGIC SETTING

At the turn of the century, following the Sino-Japanese War (1894–1895) and the Russo-Japanese War (1904–1905), Japan became a major power. Subsequently, for more than 70 years, Japanese leaders were mesmerized by the United States—by its abundant wealth and huge size, and by its capacity to change and grow.[1]

Japan thought of itself, like the United States, as a new nation. Ascending in the nineteenth century, with the ancient roots of a pure Yamato (Japanese race) in power, the emperor of Japan could trace his unbroken lineage back 2,600 years. This racial purity was, in Japanese eyes, a crucial distinction, giving its people a unique superiority. It would enable Japan to become the superpower of Asia, outstripping all its neighbors in wealth and might.[2]

In the equation of power, Western nations such as France and the Netherlands were negligible. Only Soviet Russia and, sometimes, Great Britain dominated the minds of Japanese politicians to the same extent as the United States. Emerging as a new world power with obvious expansionist tendencies, the island-nation and the United States began watching one another warily and with growing animosity.[3]

It was Japan's national spirit more than its material advantage that gave it potential, but this preoccupation with spirit and race had disturbing consequences. The Japanese hated foreigners, and their centuries of isolation were well known.[4] The visionaries of the Japanese Social Education Foundation, established in 1906, declared that the new Japanese should be a great cosmopolitan, humanitarian people.[5] The editor of the *Katsudo no Hihon*, one of a new crop of journals, suggested, "Our expansive energy should not be channeled only in the direction of Asia, but should cover the whole of mankind."[6]

This new language and ideology became common; some Japanese sought opportunities for investment in the Dutch East Indies, while others favored China and South America. Large numbers emigrated to Hawaii, the Philippines, and even Brazil. By 1907, more than 230,000 Japanese were living permanently overseas, while Japan established factories in Java and Sumatra and invested heavily in the Indies. During the quarter of a century that followed, Japan became increasingly powerful as an industrial and military nation.[7]

Japan was the first Oriental state in recent times to modernize, escaping a similar fate to that of China and other Far Eastern nations, which were eroded and overrun by Western imperialism. The Japanese felt that they had an aptitude for colonization but that the Western nations blocked them from their destiny, frustrating them for reasons that were both racial and political.[8]

Many Westerners found the Japanese repellent. In the 1920s and 1930s, when public opinion was heavily colored by racist and cultural prejudices, many in the West tended to dismiss the Japanese as "little yellow men." Only the devastating Japanese attacks on Pearl Harbor, Malaya, and the Philippines demolished the crude stereotype of the Japanese as a myopic, stunted, unmechanical people.[9]

The Japanese had a clear view of their recent history and believed they had only narrowly escaped colonial rule. Always fearful that the powerful colonial nations of the West might take advantage of any weakness, the slogan of this new era became, "Increase the nation's wealth, strengthen the army."[10] Japan devoted particular attention to building up an army and navy comparable to those of the West.

On the homefront, Japan had become an industrial nation. The pre-1914 process of industrialization had been immensely boosted by World War I, partly because of Allied contracts for munitions and a strong demand for Japanese shipbuilding, and partly because Japan's own exporters were able to enter Asian markets that the West could no longer supply.[11] Imports and exports tripled during this era, steel and cement production more than doubled, and great advances were made in the chemical and electrical industries.[12]

During the war, Japan's foreign debts were liquidated and the nation became a creditor. Japan also became a major shipbuilding nation, launching 650,000 tons in 1919, compared to 85,000 tons in 1914. As the 1919 League of Nations *World Economic Survey* showed, Japan had boosted its manufacturing production even more than that of the United States, and the continuation of that growth during the 1919–1938 period made Japan second only to the Soviet Union in overall rate of expansion.[13]

By 1938, Japan had become economically stronger than Italy and had overtaken France in all the indices of manufacturing and industrial production.[14] Despite this expansion, much of Japan's population remained in small-plot agriculture. This group suffered, not only due to rising rice imports from Taiwan and Korea, but also from the collapse of the vital silk export trade when American demand fell away in 1930.[15]

Seeking to alleviate these misfortunes by imperial expansion was always a temptation for worried and ambitious Japanese politicians. For example, the conquest of Manchuria meant economic benefits as well as military gains. For Japan's planners, expansion on the mainland of Asia was essential. It was the only means of sustaining its growing population, as only the mainland could provide the cheap raw materials needed on the islands of Japan. As Japanese confidence grew, nationalists shifted their emphasis from establishing a special position in China to pursing a divine mission—nothing less than a Japanese hegemony throughout the Far East.[16]

Japanese factories were capable of turning out great quantities of modern weapons and war equipment. It was partly this rearmament and partly the result of the exploitation of captive East Asian markets that brought industrial and economic strength. However, as Japan's industry expanded, its dependency on imported raw materials grew as well. For example, the steel industry required large amounts of pig iron and ore from China and Malaya. The Japanese islands contained few of the natural resources needed for manufacturing iron or steel, or for continued economic expansion and industrialization. Japan also lacked coal and oil.[17]

These requirements, coupled with a near-total reliance on petroleum fuels of all sorts, fostered the continued production of merchant ships to bring these resources from other countries, particularly Japan's colony in Manchuria. Japan's quest for "economic security," a self-evident good in the eyes of its fervent nationalists and military leaders, drove the nation forward, but with mixed results.[18]

At the 1919 Paris Peace Conference, Japan achieved much less than it had hoped, in particular failing to obtain a declaration of racial equality. The grant for German territories captured during the war did not fully gratify Japan's ambitions. Japanese aspirations were quite extensive, and the nation's failure to gain its objectives at Versailles upset the balance of domestic politics. It reinforced the view of the growing number of radical nationalists that Japan could expect nothing from the international community: only its own strength and resources would guarantee the future. Throughout the 1920s, every negotiation with the West seemed to involve a sacrifice of Japan's national interest.[19]

Japan's fervent nationalists saw no gains, only betrayal and loss.[20] Assassination became increasingly common in Japanese politics. Between the death of the Meiji emperor in 1912 and the outbreak of war in 1941, "six Prime Ministers were murdered, and many other politicians killed or wounded."[21] Traditional Japanese society had been based on order and hierarchy; now, both were breaking down.

The shadow over politics in Japan was the military. The army modeled the Prussian Army, and the navy, the British Royal Navy; under the Meiji constitution of 1889, both were "beyond civilian control."[22] The minister of war was always a serving military officer, as was the navy minister; thus, the armed forces operated as a state within a state. However, this inner state was no more organized and functional than the political world outside. By the late 1920s, the military was out of control. Civilians who attempted to curb its power or cut its budgets were murdered, and the high command distanced itself from the increasing turmoil.[23]

As expansion continued, differing strategic preoccupations of the army and navy fed back into foreign policy. For the army, the preoccupation was with war on the mainland, with Soviet Russia. For the navy, the dominant thought concerned lines of supply across the Pacific; its potential enemies were the American and British fleets.[24]

Over growing concern, the United States withheld precious scrap iron exports and drastically decreased trade with Japan in an effort to discourage the latter's aggression in the Far East. To the contrary, however, this strategy served only to exacerbate Japan's pursuit of its national security objectives of providing much-needed natural resources to its growing nation.[25]

Japan was still determined to continue its program of establishing a "Greater East Asia Coprosperity Sphere" through the domination of the Far East (at the expense of British and Dutch colonies). Japan hoped that this could be accomplished without a Pacific war with the United States. However, by the end of 1940 it was evident to Japanese military leaders that war was inevitable if their critical program of Asian expansion was to continue. In October 1941, after a solemn meeting with Emperor Hirohito and the Imperial Council, Japanese Prime Minister Hideki Tojo issued orders to prepare for war.[26]

NATIONAL SECURITY STRATEGY

Japan developed an enduring national security strategy as early as 1890, which was organized as follows:[27] (1) keep the home islands under the control of a central government and a unified army; (2) maintain control of the seas around Japan's islands; (3) dominate land masses abutting this area of sea control; (4) be the dominant naval power in the northwest Pacific as far south as Formosa and as far southeast as Iwo Jima; (5) secure and maintain control of access to mineral sources, in either mainland China or Southeast Asia overall, by dominating the entire western Pacific and excluding all foreign navies.[28]

NATIONAL MILITARY STRATEGY

Japan has a vast coastline and little physical depth. The country is an archipelago, consisting of a group of four large islands and numerous smaller ones, which runs about 1,200 miles from the southern tip of Kyushu to the northern tip of Hokkaido. The largest island, Honshu, is about 800 miles long, 200 miles wide at its widest point, and 70 miles at its narrowest. The four islands are divided by straits less than 30 miles wide. The terrain of all the islands consists of costal plains, with interior mountains of generally less than 5,000 feet, but rising to over 10,000 feet at certain points. An invader, once ashore, would have an easy time splitting any of the islands in two, or at least capturing the populated coastal plain. Historically, invasions of Japan from Asia have failed. Thus, Japan has never been forced to achieve its first security goal of defending the mainland.[29] This national military strategy was preeminent in the Japanese decree to secure a strategic "line of interest (not unlike a present-day country's established legal boundary, or off-shore limits)." Japan's line of interest is an outer line from which foreigners could encroach on Japan, including both ocean and land areas.[30]

Stretching from Russia's Kamchatka Peninsula to Sakhalin Peninsula, the Chinese coast from the Shantung Peninsula south to Fuzhou, then east to Formosa and the Ryuyus, and out into the Pacific to the Bonins and the rest of Micronesia, the line of interest was the foundation of Japan's second and third strategic principles: to maintain control of the seas around Japan's islands and to dominate the land masses abutting this area of sea control.[31]

In order to achieve their second goal—control of the waters around Japan—the Japanese had to invest enormous economic resources relative to the size of their economy. While remarkable gains were made in the development of merchant marine assets, the major consumer of national monies was the Japanese navy. As such, this security objective was quickly reached as Japan's navy strived to meet the national military objective of becoming the most powerful navy in the northwest Pacific.[32]

Supporting a large navy from Japan would be difficult for geographic reasons. This reality further links the dependency of the national strategic objectives to the military objectives and explains the application of supporting military security strategies. Japan is shaped like a crescent, with the two points closest to the mainland of Asia. The distance from Fukuoka, the port on southernmost Kyushu, to Pusan, Korea's southernmost port, is about 140 miles. On the Asian coast facing Japan, there are relatively few usable ports. In the north, facing Hokkaido, there are none.[33]

In the center is Vladivostok, and from there, the distance to Japan is around 5,000 miles. In the south, on the Korean Peninsula, there are several harbors, at Pusan, Wonsan, and Chongjin. However, the waters off the straits between Korea and Japan are extremely treacherous, more so even than the English Channel.[34]

The domination of the land masses abutting its areas of sea control proved to be a far more difficult security and military strategy than the Japanese planners had envisioned. Throughout the 1920s and 1930s, Japan continued the quest of its third national security objective through a program of expansion in southeast Asia. There, in the crumbling British, Dutch, and French empires, lay the oil, rubber, latex, bauxite, and other vital resources that Japan badly needed. Only the United States and Soviet Russia stood in the way, and their interference, the Japanese believed, could be checked by political alliance. Japan sought to immobilize the United States diplomatically in 1932 with the Tripartite Pact and to gain the friendship of Russia with a five-year pact of nonaggression and neutrality.[35]

Simultaneously with these diplomatic measures, the Japanese army and navy focused on their national military objectives and began to prepare actively for a general war, while laying the basis for military action in the south. In December 1940, the army ordered three divisions, then in south China, to begin training for operations in tropical areas. During the next few weeks, special studies were made on the geography, terrain, and climate of Malaya, Indochina, the Dutch Indies, Thailand, and Burma, and of the problems involved in military operations there.

Japan's quest to be the dominant naval power in the northwest Pacific made the Imperial Japanese Navy the nation's largest consumer of government funding. As a result of this increased spending, Japan was able to build heavy cruisers, which displaced close to 14,000 tons. All its major warships were fast and heavily armed. Even existing battleships had been modernized, and by the late 1930s, it was laying down the gigantic Yamato-class vessels, which were largest in the world.

The most important element of all was the extension of Japan's naval force projection via a powerful and efficient naval air service, with 3,000 aircraft and 3,500 pilots. It centered on the ten carriers, but also included some extremely efficient bomber and torpedo-carrying squadrons on land. Finally, the country also possessed the world's third largest merchant marine.[36]

Japan's actions throughout the interwar years strained relations with the United States, France, and Great Britain; the former nation continued to build its military machine and to dominate the waters surrounding the island mainland. The former allies became increasingly concerned with the size of Japan's navy, which had been built to support its third national security strategy.[37]

Ironically, Japan's fourth national security objective—being the dominant naval power in the northwest Pacific—was directly supported as a result of the Naval Armaments Treaty (signed in February 1922), in which the United States and Britain agreed to limit their navies to a total of 525,000 tons each, while Japan was allowed 315,000 tons. All parties agreed that no new construction of heavy warships (battleships, aircraft carriers, or cruisers) would be attempted for the next ten years. This treaty established a strict ratio of 5:5:3, for the numbers of ships between the major powers, which meant that for every five American and five British ships (exceeding 32,000 tons), there could be three Japanese ships of equal tonnage.[38]

As such, the treaty virtually ensured that Japan would have a much larger and more powerful Pacific navy than any other nation. Its navy concentrated in eastern Asia and in the western Pacific, but the U.S. and British fleets were to be spread throughout the waters of the world. Additionally, since the United States in the postwar stand-down did not build as many ships as it was authorized, and since Japan secretly (and illegally) built more than its allocation, the latter nation emerged in 1936 with a navy nearly two-thirds the size of the American navy.[39]

Clearly, it was the scarcity of internal industrial resources, more than any external threat, that endangered Japan. Japan's fifth, and perhaps most critical, security strategy of maintaining access to, and control of, strategic mineral resources in China and Southeast Asia led to the gravest problem of all. Japan's industry had little surplus capacity, and there was a great scarcity of raw materials. The nation was essentially in a catch-22 situation: to achieve its security objective, it needed to occupy and control terrain, but to do so, it needed to produce the required war materials and weapons.[40]

What made this strategy even more difficult to sustain was Japan's shrinking stocks of foreign currency and raw materials and its increasing dependence on imports from the disapproving Americans, British, and Dutch. Logically and inexorably, the conviction grew that Japan would have to strike south, both to isolate China and to gain a firm grip on the oil and other raw materials of Southeast Asia, the Dutch East Indies, and Borneo.[41]

Japan first had to engage in the political and military actions necessary to secure the natural resources that industrial production required. In other words, unless Japan could secure the resources necessary for its economic requirements via diplomacy, it would have to resort to war. Simply put, for Japan, war is the foundation of industry, while for the United States, industry is the foundation for war.[42]

After long negotiations, in December 1921, the United States, Great Britain, France, and Japan signed the Four Power Pacific Treaty, in which they promised to respect each other's rights and their island possessions in the Pacific. This would essentially permit Japan to control all that it had previously secured in the region, but not the much-needed resources of the Asian mainland. While the treaty temporarily supported Japan's national security strategy, it did not support its military strategy, which was designed to feed the still-growing Japanese industrial and military machine.[43]

The army in Japan traditionally stood for a course of expansion, which would make the nation the unchallenged leader in Asia. The adopted military strategy called for, among other things, the establishment of a firm position on the Asiatic continent—a euphemistic way of saying that China must be conquered; expansion into southeast Asia to secure the bases and raw materials was needed to make Japan economically strong and self-sufficient; and it was necessary to achieve the strengthening of the military forces of the nation; development of critical war industries, and improvement of air and sea transportation.[44]

Japanese military expeditions into China increased in order to gain control of an increasingly more important national security objective. In 1931 and 1932 Japan seized the Chinese provinces of Manchuria and Jehol and occupied the large city of Shanghai. In July 1937, Japanese troops attacked Chinese soldiers guarding Marco Polo Bridge near Peking, waging total war in the quest for this national security objective. Refusing to admit that this was other than what Japan termed a peacekeeping mission to restore law and order to civil war–ridden China,[45] Japan labeled the invasion the "China Incident."[46]

WEAPONS AND FORCES

In support of its national security strategy and national military strategy, Japan developed weapons and forces critical to the attainment and protection of these objectives. The military forces, land, naval, and air were equipped and organized to fulfill this strategic mission.

Land Forces

In support of its most critical national security objective—to secure and maintain control of access to mineral sources in mainland China or southeast Asia—Japan built an enormous army. In China alone, the Japanese army consisted of approximately 300,000 soldiers. Japan also had about 150,000 Manchurian and Mongolian troops under the command of Imperial Japanese officers. In addition, the military had amassed well over 2,000,000 trained reserves (still in mainland Japan), and had an estimated 150,000 soldiers dispersed throughout the rapidly expanding number of territorial garrisons.[47]

Japan's army was equipped with the most modern military weaponry available. It had ready access to manpower and could engrain the recruits into its traditions of absolute obedience and mass maximum effort. As it expanded to meet the national military objective of territorial occupation, the 24 divisions and 54 air squadrons of 1932 grew to 51 active divisions and 133 air squadrons by 1941. In addition, there were 10 depot divisions (designed as training divisions), and a large number of independent mainland brigades and garrisons, numbering around 30 divisions.[48]

The army was not strong in tanks, for which neither the terrain of the nations Japan would have to secure nor the wooden bridges of much of East Asia were suitable. It had good mobile artillery, however, and was well trained for jungle work, river crossings, and amphibious landings.[49] The army's 2,000 first-line aircraft (like the navy's) included the formidable "Zero" fighter, which was as fast and as maneuverable as anything in Europe at the time.[50]

In order to secure its national security and military strategy, Japan would have to occupy territory with its ground forces. For this reason, Japanese planners placed heavy emphasis on the training and structure of the army. In fact, the navy, with all its lethality, was essentially a support force for the ground forces. To deploy ground forces, support them at the operational level of war, and sustain them required a strategic military asset. The strategic asset that supported Japanese land forces was the third largest navy in the world; similar in composition to that of Great Britain, it was augmented by an extremely efficient carrier- and ground-based air force.[51]

Naval Forces

In December 1941, the Japanese combined fleet of some 260 major warships was the most powerful single naval force in the world.[52] Its principal strength lay in its eleven battleships and eleven aircraft carriers, which it used to project power in support of its national security and military strategy.[53]

The Japanese naval crews were maintained at full strength; they were experienced and trained in coastal operations against China, and they were mentally and physically prepared for war, wherever it might come. On the other hand, in 1941, the Western navies had crews that were only partially trained and were considerably smaller than those of Japan.[54]

The Japanese had several other technical advantages. Most critical was the larger, faster, and much more reliable "Long Lance" torpedo, which was developed especially for employment by aircraft in support of offensive operations against enemy vessels.[55] In early surface, air, and submarine actions, the Japanese torpedoes proved far superior to any others.

The Japanese warships had larger and more effective antiaircraft guns than most Allied ships. The Japanese navy had developed and perfected a nearly smokeless and flashless gunpowder. The result was naval gunfire in support of operations during darkness, which would not reveal the location of the ship to enemy counterfire observers.[56] The vast majority of Japan's warships were either new or recently modernized. The proven concept of weapons development in support of national security and military strategy was paying off.

One element of Japanese naval power that was never truly developed was that of the submarine. Japanese interest in the submarine began with the Russo-Japanese War. Both Russia and Japan purchased submarines from the United States yet neither used them in combat. This seemingly benign technology was overlooked by the Japanese naval command until late in the 1930s. Even then, the submarine was seen as a reconnaissance vessel rather than an attack asset. It was not until 1943 that the strategy for their employment was altered to include an attack role, including merchant as well as warships.[57]

Air Forces

Without fanfare, but with a single-minded determination characteristic of their nation, the Japanese seized the world's leadership in naval aviation in support of their national security objectives. With resources inferior to both the British and the Americans, and with less industrial capacity, the Japanese carried on a much greater aviation program. By the mid-1930s they had conducted more experimentation and built many more carrier planes, which were faster, farther-reaching, and more maneuverable than the Allied aircraft.[58]

An effective dive-bomber was the Aichi company's D3A1, a bat-winged Type 99 termed the "Val." It entered service in 1939, and was a rugged, reliable monoplane. The Val carried a 550–pound bomb and could fly at 241 mph at 9,000 feet. As a fleet scout, it could fly 1,250 miles, far enough for normal carrier operations.[59]

The Nakajima A6M (known as "Zero-sen" to its pilots and "Zeke" to its Allied foes), was an altogether exceptional airplane.[60] Its better than 1,200 mile range, which was twice that of contemporary American fighters, may well have been an essential precondition for the Pearl Harbor attack.[61]

CONCLUSIONS

Japan's national security strategy called for the attainment and protection of five objectives. To accomplish these objectives required its armed forces to be employed far away from the nation's border. Thus, it had to create a military force equipped to project power into neighboring states, by force if necessary; occupy and hold territories; and sustain those deployed forces. Long-reaching, sustainable, and lethal were the attributes of the Imperial Japanese Army and the unfaltering support of its navy warships and carrier-based aircraft.[62]

Japan's greatest ally is the sea, which has protected the nation for centuries and was key in the securing of its first national security objective. Japan's decision to build one of the most formidable navies ever known is logical to even casual observers. The growth of Japan's military forces matched its industrial growth, not by design, but as testimony to the theory that national security objectives of stability and economic and industrial growth are undeniably linked to military security objectives and to the development of forces and weapons to guarantee those objectives.[63]

NOTES

1. K. Latourette, *The History of Japan* (New York: Macmillan Co., 1957), ch. 5.
2. R. Overy and A. Wheatcroft, *The Road To War* (London: Macmillan, 1989), 224.
3. E. Oxford, "Prelude in the Pacific," *American History Illustrated* (September/October 1991): 53.
4. Overy and Wheatcroft, *The Road to War*, 225.
5. Ibid., 226.
6. A. Iriye, *Pacific Estrangements: The Japanese and American Expansion 1897–1911* (Cambridge: Harvard University Press, 1975), 126–129.
7. Latourette, *The History of Japan*, ch. 10.
8. H. Borton, *Japan's Modern Century* (New York: Ronald Press, 1955), ch. 2.
9. C. Thorne, *The Issue of War: States, Societies, and the Far Eastern Conflict of 1941–1945* (London: Macmillan, 1985), 25–215.
10. Overy and Wheatcroft, *The Road to War*, 229.
11. P. Kennedy, *The Rise and Fall of Great Powers: Economic Change and Military Conflict from 1500 to 2000* (New York: Random House, 1987), 97–98.
12. Ibid., 230–343.
13. Overy and Wheatcroft, *The Road to War*, 243.
14. Kennedy, *The Rise and Fall of Great Powers*, 299.
15. G. Friedman and M. Lebard, *The Coming War with Japan* (New York: St. Martin's Press, 1991), ch. 2.
16. Overy and Wheatcroft, *The Road to War*, 231.
17. I. J. Bryan, *Japan from Within* (London: T. Fisher Unwin, 1924), ch. 3.

18. J. B. Crowley, *Japan's Quest for Autonomy: National Security and Foreign Policy 1930–1958* (Princeton, NJ: Princeton University Press, 1966), 154–188.
19. Latourette, *History of Japan*, 166, 169, 175, 213.
20. R. Storry, *The Double Patriots: A Study of Japanese Nationalism* (Cambridge: Harvard University Press, 1957), 21.
21. D. James, *The Rise and Fall of the Japanese Empire* (London: Macmillan, 1951), app. 8.
22. Overy and Wheatcroft, *Road to War*, 235.
23. Bryan, *Japan from Within*, 149, 153, 156, 158.
24. Friedman and Lebard, *The Coming War*, ch. 3.
25. Ibid., 234–235.
26. Ibid., 241–243.
27. Ibid., 28.
28. Ibid., 28–29.
29. Ibid., 30–33.
30. M. Barnhart, *Japan Prepares for Total War* (Ithaca, NY: Cornell University Press, 1987), 22.
31. Friedman and Lebard, *The Coming War*, 30.
32. Ibid., 30–31.
33. G. Trewartha, *Japan: A Physical, Cultural, and Regional Geography* (Madison: University of Wisconsin Press, 1945), 52–55.
34. H. Okazaki, *A Grand Strategy for Japanese Defense* (Washington, DC: n.p., 1986), 4; extracted from G. Friedman and M. Lebard, *The Coming War with Japan* (New York: St. Martin's Press, 1991), 27–28.
35. Friedman and Lebard, *The Coming War*, ch. 3.
36. Latourette, *The History of Japan*, 182, 186, 211, 217–218.
37. Ibid.
38. Friedman and Lebard, *The Coming War*, ch. 4.
39. T. N. Depuy, *The Naval War in the Pacific: The Rising Sun of Nippon* (New York: Franklin Watts, 1973), 58.
40. Friedman and Lebard, *The Coming War*, 77–85.
41. Ibid., 83–85.
42. Ibid., 31.
43. Bryan, *Japan from Within*, 158–159.
44. T. N. Depuy, *Asiatic Land Battles: The Expansion of Japan in Asia* (New York: Franklin Watts, 1973), 26–35.
45. Ibid., 52.
46. Ibid., 11.
47. Hisahiko, *Grand Strategy*, 30–37.
48. Dupuy, *Asiatic Land Battles*, 34–39.
49. On the mobile artillery, see Bryan, *Japan from Within*, 162–163.
50. J. Horikoshi, *The Eagles of Mitsubishi: The Story of the Zero Fighter* (Tokyo: Kappa Books, 1970); reprint, trans. Shojiro Shindo and Harold Wantiez, Seattle: University of Washington Press, 1981), 29.
51. Bryan, *Japan from Within*, ch. 11.
52. Ibid., p. 220.
53. Ibid., p. 176.
54. Dupuy, *The Naval War in the Pacific*, 13.

55. This same torpedo would be modified for maximum effect in the relatively shallow waters of Pearl Harbor.

56. Dupuy, *The Naval War in the Pacific*, 15.

57. D. Carpenter and N. Polmar, eds., *Submarines of the Imperial Japanese Navy* (Annapolis, MD: Naval Institute Press, 1987), 24.

58. Horikoshi, *Eagles of Mitsubishi*, chs. 5, 6.

59. Ibid.

60. J. H. Belote and W. H. Belote, *Titans of The Seas: Development and Operations of Japanese and American Carrier Task Forces during World War II* (New York: Harper and Row, 1975), 23.

61. Ibid., 95.

62. Latourette, *The History of Japan*, ch. 10.

63. Ibid.

Chapter 4

Historical Model of the Theory: The United States in the Cold War Era, 1945–1989

This chapter illustrates the linkage among the strategic situation in which the United States found itself in the Cold War era, its national security strategy, national military strategy, and the kinds of weapons and forces that the nation develops to support power projection and achieve those strategic objectives.

THE STRATEGIC SETTING

The period since the end of World War II has been unique in the American experience. During this time the United States was thrust from the role of supporting actor to that of a leader on the world stage and was pitted against the Soviet Union, which was equally unprepared for the spotlight. The ruins of the old international order—characterized by the shrinking influence of the traditional European powers and the crumbling of colonial empires, principally in Africa and Asia—formed the backdrop of this drama. Over the entire scene was cast the increasing pall of the destructive capacity of the conventional and nuclear arsenals possessed by the superpowers.

History did not prepare Americans well for this new role, especially since world leadership was now defined in terms of which nation was most secure from the threats of others. Before 1945, there had never been anything resembling a comprehensive U.S. national security strategy; the buffers of two broad oceans and friendly, weak neighbors had rendered such a strategy unnecessary.[1] However, the defense-in-depth provided historically by the oceans that separated the United States from the zone of conflict vanished as a result of technological innovation. Thus, the nation became vulnerable to massive attack on its homeland at a time when the power balance in Europe could no longer be maintained by a combination of the European states themselves.[2]

The world had changed in many ways, three of which stand out in this discussion. First, power in the international system became *bipolar,* with only the United States and the Soviet Union retaining the capacity to significantly influence events. Each nation was drawn, whether willingly or unwillingly, into power vacuums around the globe. In the process, Americans began to regard their national security interests and objectives, including those that might potentially be defended with force, as global in nature.

Second, most nations in the world, including the United States and the Soviet Union, signed the United Nations charter renouncing war as an instrument of policy. Unfortunately, however, the result was not a positive condition of peace but an era in which force continued to be applied under various guises. This "violent peace" blurred the traditional distinctions between war and peace.[3]

Third, the nuclear age began. By unleashing atomic power, humankind created a means of self-immolation and thereby altered the basic rules for defense and war. Again, as conventional and nuclear arsenals achieved their current deadly levels, the result was to restrain the nations that possessed them. The informal rule that had emerged was that the United States and the Soviet Union could no longer afford to confront one another in ways that could lead to conventional conflict, which now had the potential to escalate to nuclear war.[4]

Fear of both conventional and atomic exchange thus governed the behavior of the superpowers toward one another. As a result, something of a paradox has emerged: Americans had consciously driven technology into a dimension of lethality that defied explanation and comprehension. Instead of guaranteeing the security of nations via the technological evolution of weapons, traditional deterrence has become a global religion of fear—it has changed from a stance of national self-assertion to one of total annihilation.

The international (or external) threats to the United States in this era could be categorized as either military or economic. The primary military threat to the United States in this period had been the Soviet Union. Beginning in the 1960s, the Soviets undertook a rapid expansion of their military capabilities.[5]

The economic threat facing the United States from 1945 to the 1960s was focused initially on the rebuilding of Europe and the island nations of Japan, Borneo, and the Philippines.[6] Involvement in the Korean and Vietnam Wars was a national economic drain but did not harm the private and corporate sectors. In time of war, the gross national product (GNP) of the United States is traditionally higher than it is during times of peace. It is not the element of peace that degrades the GNP; rather, it is the repercussion of the process that follows the end of war—the economizing of the defense force structure and budget. Every state in the union has some defense-related industry, and the procurement of weapons systems for the U.S. Department of Defense (DoD) is almost exclusively an American enterprise. As such, cuts in defense spending result in a degradation of the domestic economic situation. Increases in defense spending are positive, as they pertain to the latter; however, they do extract a toll on global economic well-being.[7]

If one excludes the incremental costs of the war in Southeast Asia, by 1968 the Soviets had forged ahead of the United States in the amount of money allocated annually to defense. Moreover, because the Pentagon was forced to expend a large percentage of its budget, first for the prosecution of the war in Southeast Asia and then for paying additional personnel costs caused by the changeover to the all-volunteer force (AVF). "The Kremlin began to outstrip the United States in outlays for a wide spectrum of military capabilities."[8]

In the 1980s, increasing U.S. dependance (and that of its allies) on foreign sources of oil and other critical industrial materials weakened U.S. economic strength and strategic deterrent capabilities. The nation imported over half its petroleum needs and had virtually excised the mining of its own abundant sources of coal. Any sudden disruption of the oil flow could bring Western economies crashing down. Even threats of such cutoffs hinder U.S. ability to pursue freely its national security objectives in such places as the Middle East and Africa.[9] In 1979, for example, pressure from Saudi Arabia complicated U.S. efforts to arrange a settlement between the Egyptians and the Israelis, and threats from oil-rich Nigeria became a factor in U.S. policy toward Rhodesia and South Africa.

World War II reaffirmed the fact that the United States should not pretend to be immune from global turmoil and gave birth to the notion of the United States as a superpower. Along with Britain and the Soviet Union, the United States was expected to assume the responsibility for reconstructing and preserving a new world order so as to avoid the horrors of yet another total war.[10]

The Truman Doctrine and the Marshall Plan of 1947, the articulation of "containment" and the beginning of peacetime alliances such as the North Atlantic Treaty Organization (NATO), and the U.S. response to the attack in Korea in 1950 all represented major milestones in the emergence of the United States as the Western superpower.[11] It was now able to lead a coalition of states in containing the opposing superpower and its bloc of satellites, which had a quite different ideological and political idea of what the postwar world order ought to be.

While the Rio Pact (1947) represented a long-standing U.S. concern for hemispheric security, which was rooted in the 1923 Monroe Doctrine, the formation of NATO in 1949 reflected an unprecedented peacetime commitment.[12] Additional collective security treaties included the Australia, New Zealand, U.S. pact (ANZUS, 1951); the Southeast Asia pact (SEATO, 1954); the Central Treaty Organization (CENTO, 1955); and bilateral defense agreements with the Philippines (1951), South Korea (1953), Taiwan (1954), and Japan (1960).[13] These were complemented by a variety of agreements on bilateral defense cooperation with friendly states around the world, forming the structure for both global military commitments and a global military presence in support of U.S. national security.

NATIONAL SECURITY STRATEGY

By the end of World War II, a new term, *national security*, was finding increasing acceptance in policy circles in Washington. It was a much more sweeping concept than the ideas of diplomacy and defense, which it replaced, and reflected the technological, political, and geographic forces that had emerged during World War II. It implied that the safety of the nation was dependant on a wide, interrelated range of foreign policy issues, which proved much more complex and diverse than anything the nation had previously faced.[14]

In the late 1940s and well into the early 1950s, U.S. national security strategy was conceived under the basic idea that the United States should sit back and allow a hostile power like the Soviet Union to acquire a massive nuclear arsenal and that a much more active and positive policy had to be seriously considered, a belief that was surprisingly widespread.[15] In response to this demand, three security policies, containment, deterrence, and economic security, emerged over the next decade and remained operative throughout the Cold War era.

During the first five years of the nuclear age, containment and national security came to dominate America's thinking about foreign policy. As such, the fundamental goals that would determine foreign policy through the Cold War era were now in place. Soviet power had to be contained, deterrence of general war must be successful, and the Soviet ideology must be transformed and quelled.[16]

These goals remained stable throughout the Cold War period. The different strategies and policies that the various administrations from Harry Truman, through Ronald Reagan selected during their tenures to enforce these goals varied widely. These divergent strategies and policies were the result of differences in the interpretation of strategic goals and the acceptance of renovating documents like NSC-68.[17] Regardless, "containment—the bottling up of Soviet expansionism—was the foundation of U.S. security policy."[18] Containment was a desired condition in U.S.-

Soviet relations and was considered a stable policy aimed at a general balance of power in the world.[19]

Deterrence—another security policy—had been fully incorporated into defense planning strategy by 1948. With the advent of atomic weapons, deterrence of nuclear war became the fundamental objective of U.S. national security strategy. First, deterrence required that the United States have the capability to carry out a true, countervailing strategy.[20] The second objective of deterrence was to keep open the air and sea lines of communication between the United States and its allies and trading partners. The third objective was to provide an international environment in which democratic development is encouraged and free access to overseas markets for American businesses is maintained. Fourth, the United States desired to conclude any hostilities that might occur with a minimum loss of U.S. life and on terms favorable to itself and its allies.[21] The fundamentals of deterrence and the synergy of atomic weapons and strategy were understood in both the military and diplomatic communities.[22] Simply put, guaranteed retaliation became the key functional element to deterrence.

The third defense policy, economic security, was another element of Cold War security strategy. The objective was to maintain an environment in which U.S. companies had access to overseas markets for trade and investment. This objective arose from two sources: the belief that unrestricted trade would do for the international system what laissez-faire economics had done for this nation, and second, that the U.S. standard of living depends on the ability to secure certain raw materials, invest overseas, and have access to world markets.

There is no doubt that the U.S. economic system and its underlying philosophy play a role in determining U.S. national strategy. For example, the United States would not stand idly by while a raw material cartel undermined a country's economy. Admittedly, there are some exceptions, whereby the United States has often acted contrary to its own best economic interests. For example, it has consistently supported Israel despite becoming increasingly dependent on Arab oil. However, the support is intended to counter Soviet influence in the Middle East, this is an example of U.S.-supported deterrence intended to facilitate the U.S. containment strategy.[23]

Even during the oil embargo of the 1970s, the United States continued to send massive amounts of aid to Israel. For nearly three decades after World War II, the U.S. government severely restricted trade with the Soviet Union, mainland China, and Eastern Europe, despite great pressures from the "collective corporate" (corporate sector) to access these markets.[24]

Such a definition of national interests—the kind of core interests that merit defense, if necessary, by armed force—led to a number of specific national security objectives which, in the following section, are discussed in relation to national military strategy. Such objectives generally include the following: (1) deterring aggression against the United States, its forces, and its allies; (2) should deterrence fail, defeating armed aggression and terminating conflict on favorable terms at the lowest possible level of conflict; (3) securing the United States and its allies from coercion or intimidation by hostile forces; (4) facilitating the ability of allies and friends to defend themselves against aggression, coercion, subversion, insurgency, and terrorism; (5) ensuring U.S. access to critical resources, markets, the oceans, and space; (6) where possible, reducing the dominant threat of Soviet Military power by a variety of means short of the direct use of military force (such as political influence, restrictions on access to military technology, and agreements to control, limit, or reduce armaments).[25]

National Military Strategy

Deterring aggression against the nation and its forces is the first and most critical security goal of the United States. As such, defense of the American homeland is the first priority of U.S. military strategy. Because strategic nuclear war with the Soviets represented the only level of conflict that would immediately threaten the survival of the United States, the major military objective has been the deterrence of strategic nuclear conflict. The cost of failing to deter strategic nuclear war is so high that U.S. security planners rightly gave priority to the precise management of forces to ensure strategic deterrence.[26]

Another national security interest during the Cold War era was the security of U.S. allies; commitment to this resolve is indicated by defense strategies supported by such documents and alliances as the Truman Doctrine, the Marshall Plan, the Berlin airlift, and formation of the North Atlantic Treaty Organization. The policy statements, force development plans, and weapons employment decisions made during this era of assured mutual destruction reflected the U.S. desire to achieve its security objectives. Three decisions were made to help influence events. First, the United States would not develop an antiballistic missile defense, and a 1972 U.S.-Soviet agreement to limit employment of antiballistic missiles seemed to enshrine both nations in the declared concept of mutual destruction as their one mutual interest.[27]

Second, the United States elected not to build new strategic forces to match the Soviet buildup in intercontinental missiles, which was well under way in the early 1970s. Instead, from its leading position, the United States chose to accept parity or sufficiency, to limit the number of strategic launchers on both sides through arms control agreements, and to ensure the deterrent values of its strategic forces through the modernization of each leg of the force: land, air, and maritime. Finally, and simultaneously, the decision was made not to restrict the deployment of MIR Ved systems (systems using multiple warheads carried by one missile or delivery platform). Thus, while the size of the U.S. force appeared to remain the same, the number of warheads that could be delivered by that force—and number of targets covered—was significantly increased.[28]

These decisions were not made irrationally. Rather, they were made as part of a total defense strategy, comprised of elements of diplomacy, deterrent capability, and force structure. This strategy supported the second national security interest of ensuring that in the event deterrence failed, the U.S. and its allies could defeat any armed aggression and terminate the conflict on favorable terms at the lowest possible level of conflict. If it was clear that the United States was willing to negotiate quantitative weapons issues in order to assure global security—which was U.S. defense strategy—then it was also clear to the United States that the weapons deployed would have to be technologically superior to the threats in order to continue a policy of deterrence and assured mutual destruction in support of U.S. defense strategy.

The next two security objectives can be combined for an analysis of military strategy applications. These national security objectives were to secure the United States and its allies from coercion or intimidation by hostile forces and facilitate the ability of allies and friends to defend themselves against aggression, coercion, subversion, insurgency, and terrorism.

Generally speaking, the national command authority of this era considered that there were three distinct threats to the security and economic well-being of the United States and its allies. The first was the challenge posed by the Soviet Union and its Warsaw Pact allies. Second was the various forms of instability that existed in the Third World and could be exploited by outside powers such as the Soviet Union. Third was the potential disruption of the global economy to such a degree that it could seriously jeopardize the economic health and well-being of the United States and its allies.[29]

In the 1960s and 1970s, it was determined that something more than a policy of "massive retaliation" was needed to enforce the latter two security objectives—specifically, the strategy of facilitating the ability of U.S. allies to defend themselves. U.S. defense policy planners had to exercise great diplomatic care not to directly sponsor strategic nuclear proliferation in an effort to secure these two security objectives. As such, defense policy planners designed a policy of flexible response. This policy dictated that the United States have sufficient nuclear weapons to fight theater wars or engage in a strategic exchange. The policy also made provisions for adequate conventional forces to fight either major wars or smaller, brushfire wars; such forces would be forward-deployed in the theater and were considered the most credible deterrent in support of this national interest.[30]

The objective of ensuring U.S. access to critical resources, markets, the oceans, and space was interpreted in the national military strategy as taking the steps necessary to protect the sea, land, and air lines of communication (LOCs) that support U.S. interests. This includes denying access to these LOCs by any force whose purpose is adverse to the interests of the United States. To achieve this objective, the United States would employ military forces and resources necessary to accomplish this mission, as may be directed.

In order to achieve the sixth security objective—where possible, to reduce the dominant threat of Soviet Military power by a variety of means short of the direct use of military force (e.g., political influence; restrictions on access to military technology; and agreements to control, limit, or reduce armaments), the United States was willing to sacrifice the analytical basis of force planning in order to achieve a perceptual strategic parity with the Soviet Union.[31]

This strategic parity was achieved via a number of means. For example, while U.S. diplomats negotiated strategic weapons issues with the Soviets, the United States continually updated the technological sophistication of its existing strategic nuclear weapons. The "Peacekeeper" (MX) missile was designed and employed as the U.S.-Soviet dialogue continued. The MX modernized and replaced the ten-year-old techology of the intercontinental ballistic missile (ICBM) and its contributions emanated from its combination of throw-weight and accuracy, equating to a hard target kill capability.[32] Another issue relevant to the MX involved discussions with the Soviets regarding the number of fixed missile silos on the European continent. As discussions evolved around this issue, the MX was adapted from a fixed to a mobile system, thus defeating the limitations imposed by the fixed silo.

It could not be shown conclusively that a mobile MX deployment would greatly enhance deterrence, and 50 percent of the U.S. ICBM capability remained in fixed silos. While this percentage remained fixed, the perception portrayed to the Soviets was that mobile MX deployment was not negotiable. The U.S. concerns were realistic; for reasons involving arms limitations and budgetary constraints, the enormous retaliatory potential of the Soviet SLBM and bomber forces remained unaltered. The ultimate rejection of mobile basing for MX missiles was due, in part, to the Soviet agreement to accommodate strategic weapons reductions. Thus, if the decision to base the "Peacekeeper" in "Minuteman" silos begot some Soviet numeric missile reduction, it may be said that the political concept of perceived equality has triumphed over an analytical measure of deterrent value.[33]

In Defense Secretary Caspar Weinberger's final *Annual Report to the Congress*, covering fiscal year 1988, U.S. national interests were framed as follows:

> U.S. National Interests encompass both broad ideals and specific security needs. America's paramount national interests are peace, freedom, and prosperity for ourselves and our allies and our friends, and for others around the world. We seek an international order that encourages self-determination, democratic institutions, economic development, and human rights. We endorse the open exchange of ideas and other measures to encourage understanding between peoples.
>
> More specifically, we maintain our steadfast concern for the security and well-being of our allies and other nations friendly to our interests. We oppose the expansion of influence, control, or territory by nations opposed to freedom and other fundamental ideals shared by America and its allies. The peaceful existence and prosperity of democracies is the core United States interest.[34]

What is critical about this statement is that U.S. national interests are defined principally in terms of the defense of U.S. ideals and those who share them. It is not just a question of territorial defense, but a basic recognition that the "American way of life," however defined, is linked to an international order, which must be defended diplomatically if possible, but if necessary, militarily.[35]

WEAPONS AND FORCES

The United States is essentially an island nation. Over 60 percent of its borders consist of shoreline. Although it has neighbors directly on its northern and southern borders, these nations have never posed a threat to the U.S. national security. Moreover, the orientation of America has always been more east and west than north and south. Its potential opponents lie overseas, and the majority of its trade has always been overseas as well, and not overland.[36]

As the century progressed, military force and preparedness became more important elements in the national sense of well-being. Geographically remote areas and issues took on more importance than before the wars, and military and foreign policy became much more closely intertwined.

At the close of World War II, U.S. armed forces numbered around 12 million soldiers, sailors, marines, and airmen. The U.S. Army Air Corps possessed 218 operational strategic and tactical combat groups and the world's only strategic nuclear weapons arsenal. The American navy consisted of approximately 1,200 combat ships, 40,000 aircraft, and 500,000 marines, making it the world's most effective projection force. Finally, the army had approximately 6 million troops organized into over 100 combat-ready divisions.[37] All this was to change dramatically as the national security strategy was translated into a weapons and force structure designed to meet its changing needs.

First to change in this Cold War era were the descriptive categories in which U.S. defense forces were placed. Defense forces were no longer referred to solely as land, sea, or air; rather, they were spoken of in terms of force mix, meaning a combination of land, air, and sea forces specifically tailored to more efficiently achieve defense strategies supporting the national military strategy. The military forces of the United States during this period were broken down into three categories: conventional, strategic, and theater nuclear, and were both stationed on U.S. soil and forward-deployed in many different parts of the globe.[38] In Chapter 3 the military forces of Japan were discussed in terms of land, sea, and air forces. As further testimony to the fact that force structure is linked to defense policy and national strategy, Chapters 4 and 5 discuss force structure in terms of the three categories inherent to the Cold War forces.

Conventional Forces

Starting at the beginning of the Cold War, the United States has stationed large numbers of military forces overseas, primarily in Europe and the Far East. In Central Europe (mostly in the Federal Republic of Germany), the United States maintained up to 245,000 soldiers and air forces staffing combat and support elements capable of waging full-scale conventional and nuclear war. These forces were positioned in support of the first and second U.S. national security objectives of deterring aggression against the United States, its forces, and its allies and of defeating any armed aggression and terminating the conflict on favorable terms and at the lowest level of conflict. These forces included the equivalent of 5⅔ army divisions and 8 air force fighter-attack wings.

In continued support of the first and second security objectives and their accompanying defense strategies, the United States deployed the Sixth Fleet to the Mediterranean. The fleet was the dominant allied naval force in the region, and like its land and air counterparts, it was a force capable of full-scale offensive and defensive operations, including amphibious operations. Throughout the western and northern Mediterranean littoral, the United States maintained in excess of 100 military facilities engaged in diverse missions ranging from logistics supporting conventional force structures to intelligence collection activities.

Because the principal national interests of the United States lay in Western Europe, the stationing of U.S. troops was not as extensive in the Far East.[39] In terms of the numbers of personnel assigned, the figures were less significant, while the geographic area covered was considerably larger than that of Western Europe. The air force maintained major bases in the Philippines, Okinawa, South Korea, Guam, and the Japanese main islands. Three operating tactical fighter-attack wings were based there: one in the Philippines and two in South Korea. Because of the expanse of the region, conventional naval and air forces were required to project power and support U.S. security strategy.

The primary U.S. Army presence in the Far East was the Second Infantry Division in South Korea. This conventional force was positioned in this theater in support of the third and fourth U.S. national security objectives: to secure the United States and its allies from coercion or intimidation by hostile forces and to facilitate the ability of its allies to defend themselves. Another conventional force sustaining the attainment of this goal was a marine amphibious force (MAF), which included a division, a tactical air wing, and combat support units. Rounding out the U.S. force structure in the Western Pacific was the U.S. Seventh Fleet. Like its opposite number in the Mediterranean, this fleet represented the cutting edge of conventional U.S. forces and was on call to perform the full range of operations in support of U.S. security and military strategies.

Seen as the forces necessary to protect vital U.S. interests throughout the globe, forward-deployed land and air forces in the European and Asian theater used technology to develop and employ weapons fitting their deterrent roles. In the European theater, forces and weapons were designed to thwart the Soviet offensive on the plains and in the cities of Europe. As such, mechanized ground forces using the M113 armored personnel carrier and, eventually, the high-tech Bradley Fighting Vehicle were the principal combat forces. Infantry was supported by armor forces using the faster, heavier, M1 Abrams tank and new, longer-reaching artillery and surface-to-surface missiles. Attack helicopters with long-standoff antitank capabilities were designed and employed in support of forward-presence, deterrent strategies. The air force, in a close air-support role, designed the antitank A10 "Warthog" aircraft to destroy Soviet tanks. All this was in support of the national security and military strategies of protecting U.S. forces and its allies against Soviet aggression on the plains of Eastern Europe.

Strategic Forces

Since the early 1960s, the U.S. strategic offensive capability was built around three complementary systems for the delivery of nuclear weapons. Referred to as the triad, these systems were land-based intercontinental ballistic missiles (ICBMs), sea-based and-launched ballistic missiles (SLBMs), and manned bombers.[40]

Two of the legs of the Triad, the ICBMs and the manned bombers, comprised the combat elements of the U.S. Strategic Air Command. They were under a single commander in chief, who reported to the secretary of the defense through the Joint Chiefs of Staff. The ICBM force consisted of 450 Minuteman IIs, 550 Minuteman IIIs, and 54 Titan IIs, all based within the continental United States. A mobile ICBM (the MX), which was designed as a replacement for increasingly vulnerable, fixed-position ICBMs, was developed but received only limited deployment, as it was used as a deterrent bargaining tool.[41]

The bomber force included two wings of FB-111s and seventeen wings of B-52s in the continental United States, plus one B-52 wing in Guam. About half the B-52 force was expected to be armed with the cruise missile beginning in the early 1980s.[42]

The navy's fleet ballistic missile submarines were the third leg of the triad. These "Polaris" and "Poseidon" submarines were gradually replaced by the "Trident" submarines, the first of which was launched in 1979. The older Poseidon ships were modified to join the Trident ships in carrying the 7,400–kilometer range Trident I missile. After the Trident II missile was developed, in the late 1980s, the Trident ships were backfitted with the improved missile.[43]

Finally, the United States possessed a small strategic defense force. During the later part of the Cold War, it consisted of just 141 active and 190 reserve fighter-interceptor aircraft based in the continental United States and a deactivated "Safeguard" anti-ballistic missile complex at Grand Forks, North Dakota.[44]

Theater Nuclear Forces

Near the end of the Cold War era, the United States possessed about 13,000 theater or tactical nuclear weapons. These consisted of battlefield systems for use by the artillery, bombs carried by aircraft, and fleet-defensive systems. Approximately 7,000 of these weapons were dedicated to the European theater. The majority were in the hands of U.S. forces, while the remainder were deployed with the forces of NATO allies, subject to American control. Another 3,000 were afloat or deployed in the Pacific, while the remaining weapons were based in the United States.[45]

Following is a generic discussion of specific force structures and the reasons for their design in support of national security strategy and military strategy.

Land Forces

In accordance with the national security strategy of deterrence, the army strategy necessary to comply with the priorities of maximum deterrence, and maximum efficacy should the deterrent fail, dictates the shape of U.S. land forces as something vastly different from the shape that evolved from army history and accepted doctrine. The whole concept of army strategy based on mobilization and a period on which to expand the peacetime nucleus had to be discarded.

The biggest contribution the U.S. Army made to the deterrent of total war was, without question, its forward presence, supported by the tactile air forces of NATO. In order for ground forces to effectively serve as a deterrent, they must be forward-deployed into friendly states, and preferably those bordering hostile states.

Basing rights, sustainment infrastructure, and force weapons tailoring, were all critical aspects of the evolution of U.S. ground forces in support of national strategy. Traditional "return to base armies" became forward-deployed armies designed to exist indefinitely while bordering on a potentially hostile state.

Naval Forces

One could argue quite convincingly that, in the age of "intercontinental ballistic missiles and huge air transports, naval strategy is no longer important and that geography should no longer be a determinant of U.S. strategy."[46] However, traditions die hard. When confronted with a crisis, policy makers almost instinctively turn toward naval forces. In over 80 percent of the cases in which military forces were employed for political purposes in the post–World War II period, naval units were the chosen instrument.[47]

For the greater part of this century, the United States regarded a naval strategy as the best hope of providing for its national defense. For example, during the 1930s, the U.S. Navy was among the three strongest in the world, while the U.S. Army ranked seventeenth. The naval strategy, which owed its intellectual rigor to Alfred Thayer Mahan, dictated that the United States control the seas in order to ensure prosperity in peacetime and victory in war. Controlling the seas required a large navy and suitable overseas bases.[48] This naval strategy meant that even when it demobilized its navy between the Spanish-American War and World War I and between World Wars I and II, the United States maintained a comparatively large force.

The United States adopted a language supportive of its security strategy. Accepted was the term *maritime* in lieu of *naval* because *naval* introduces the notion of naval warships, whereas in the Cold War period, policy makers desired to focus more on the function rather than on the instrument by which the function is performed. Simply put, an armed merchant vessel protecting itself against aircraft patrolling a sea route or striking at a naval warship or merchant ship is as much a maritime force as a battleship, yet could not correctly be designated by the word *naval*.[49]

A result of the evolution of the strategic deterrence is the required expansion of NATO maritime forces. More ships needed to be in more places at one time, and had to carry much larger payloads of long-reaching nuclear and conventional weapons. Smaller ships were developed to meet this requirement, and the traditional surface fleet was heavily augmented by the new, high-tech nuclear submarines.[50]

Submarines were used in countersurface roles almost exclusively until the advent of nuclear weapons. They were used in support of deterrent strategies to transport multiple warhead systems deep into the international waters bordering the threat. Cloaked in secrecy, capable of remaining under the surface almost indefinitely, and armed with weapons targeted at the industrial and political heart of the threat states, the submarine developed into one of the most valuable deterrent weapons systems available.[51]

The new maritime forces and their weapons also brought new considerations into overseas bases, and much like the new fleets themselves, the future vulnerability and usefulness of the forces depended heavily on the inherent linkage of NATO forces in support of the deterrent strategy. Expansive basing rights in numerous allied ports are required to keep the maritime strategy functioning. Regardless of the loiter capabilities of surface vessels, the maintenance of extensive global maritime coverage requires an interlocking net of friendly ports. Historically absent, this need was met in modern times as a result of the diplomatic aspects of deterrent strategy, which reestablished the dovetailing of national strategy, whether military or security, with the weapons and forces required to sustain it.

Air Forces

Should total war occur as a result of failed deterrence, the United States Air Force as well as its allies have developed weapons systems fully supportive of that role. The ideal weapons with which to wage total war, these are the ultimate in their respective roles because they were developed for their deterrent value, which theoretically makes them better performers of their war tasks.

New weapons developed for their deterrent role will increase the importance of the primary roles of the air force: on the one hand, with weapons of mass destruction and long-range bombers, and ballistic rockets (the deterrent value and destructive power of the offensive), and on the other, with defensive guided missiles (the protective power of the defense).

Ground-to-air guided missiles were developed as a superior replacement to medium and heavy antiaircraft weapons, as a result of extended-range bomber development as well as developments in supersonic fighters and medium-range bombers. The strategy of missile deployment and employment created new considerations unique to the national security strategy.

The dominating factors are the range and effective height of the missile and some form of adequate warning system. Perhaps the most difficult problem in the strategy of strategic missile defence is the technical development: deciding the characteristics of the weapons that will prove them superior, at not too great a cost, relative to the enemy's offensive weapons.

CONCLUSIONS

American defense policy throughout the Cold War period was designed to secure the United States in its role as a world power, a role that the nation inherited as the clear legacy of World War II. All the elements that shape that global role, however, are changing, whether in superpower relations, the U.S. role in Europe or military presence in Asia, the growing challenges of nonterritorial security threats, or the resources likely to be able to sustain that role.

However, in focusing on the elements of a global defense posture and the problems associated with sustaining the forces necessary to secure U.S. interests and commitments, it is easily forgotten that the United States is a world power, and not simply because of its military prowess or because of the unique role derived from being a nuclear superpower. That global position is, fundamentally, derived from the political, economic, social, and philosophical influence of the United States, which has, over the past four decades and more, tolerated a degree of elasticity in the structure of U.S. military forces and the form in which the nation has met its commitments. The United States and its allies have been able to accommodate the fluctuations in U.S. defense posture also because the overriding purpose of that defense posture is deterrence. Nonetheless, credible military power is vital both to deterrence and, fundamentally, to the interests and commitments of a world power.

NOTES

1. D. M. Snow, *National Security: Enduring Problems of U.S. Defense Policy* (New York: St. Martin's Press, 1987), 71.

2. R. L. Pfaltzgraff, Jr., *National Security: Ethics, Strategy, and Politics—A Layman's Primer* (New York: Pergamon and Brassey's, 1988), 24–25.

3. Ibid., 32–34.

4. Forgive my now-dated analysis using the Soviet Union and the United States. It is arguable that I should redefine the threat in more accurate terms, given that the USSR is no longer a collective entity. However, it is not the pur-

pose of this chapter to define enemies; rather, it is to suggest alternative weaponry (*nonlethal*) in support of national security, whatever the threat.

5. W. R. Kitner and R. B. Foster, eds., *National Strategy in a Decade of Change: An Emerging U.S. Policy* (London: Lexington Books, 1973), 171–177.

6. G. B. de Huszar, *National Strategy in an Age of Revolutions* (New York: Praeger, 1958), 134.

7. Ibid., 72–73.

8. U.S. Central Intelligence Agency, "Estimated Soviet Defense Spending: Trends and Prospects," June 1978, 1.

9. In the eyes of the American people, the Persian Gulf War will continue to be perceived as a war designed to trade blood for oil. Regardless, however, the U.S. National Command Authority and Congress will never admit to that; for them, we went to war in defense of liberal democracy—to thwart aggression. However, had Saddam Hussein kept Kuwait and subsequently captured the oil fields to the south, in Saudi Arabia, he would have controlled some 68 percent of the world's oil.

10. Kitner and Foster, *National Strategy in a Decade of Change*, 221.

11. Ibid., 234–238.

12. J. Bayles et al., eds., *Contemporary Strategy: Theories and Policies* (New York: Holmes and Meier, 1977), 37–40.

13. Ibid., 42.

14. Ibid., 45–46.

15. In September 1954, a Gallup poll asked: "Some people say we should go to war against Russia now while we still have an advantage in atomic and hydrogen weapons. Do you agree or disagree with this point of view?" Thirteen percent of the sample agreed, 76 percent disagreed, and 11 percent had no opinion. Similarly, in July 1950 (right after the outbreak of the Korean War), 15 percent of a Gallup sample thought the United States should declare war on Russia immediately. H. G. Erkine, "The Polls: Atomic Weapons and Nuclear Energy," *Public Opinion Quarterly* 27 (1963): 177; G. Gallup, *The Gallup Poll: Public Opinion 1935–1971* (New York, 1972), 1: 930.

16. Gallup, *The Gallup Poll*, 22–23.

17. D. J. Sullivan, "Dealing with the Soviets," in *American Defense Policy*, 6th ed. (Baltimore, MD: Johns Hopkins University Press, 1990), 165–172.

18. Ibid., 165.

19. Ibid., p. 173.

20. J. P. Murray and R. Viotti, *The Defense Policies of Nations: A Comparative Study* (Baltimore, MD: John Hopkins University Press, 1982), p. 68; see also H. Brown, "Annual Defense Department Report, FY 1980," 25 January 1979, 1–9.

21. Extract from Murray and Viotti, *Defense Policies of Nations*, 56–57; drawn from the Annual Defense Department Reports (FY 1978, 1979, 1980) and speeches from President Jimmy Carter at Wake Forest University on 17 March 1978, and at Georgia Tech University on 20 February 1979.

22. Ibid., 171.

23. W. J. Levy, *Oil Strategy and Politics, 1941–1981* (Boulder, CO: Westview Press, 1982), 269–272.

24. *Collective corporate* is a term I coined.

25. Kitner and Foster, *National Strategy in a Decade of Change*, 104–105; see also A. N. Sabrosky's *Polarity and War: The Changing Structure of International Conflict* (London: Westview Press, 1985), 191–209, and S. Foerster and F. N. Wright, eds., *American Defense Policy* (Baltimore, MD: Johns Hopkins University Press, 1982), 285–288.

26. H. D. Graves, *U.S. Capabilities for Military Intervention: U.S. Policy and Low Intensity Conflict* (New Brunswick, NJ: Transaction Books, 1981), 69.

27. Foerster and Wright, *American Defense Policy*, 349–351.

28. Kitner and Foster, *National Strategy*, 171–179.

29. L. J. Korb, *Defense Policy of the United States* (Baltimore, MD: Johns Hopkins University Press, 1986), 51–52.

30. Ibid., 56–57.

31. R. Haffa, Jr., *Force Planning: Bridge between Doctrine and Forces* (Baltimore, MD: Johns Hopkins University Press, 1989), 496–497.

32. Ibid., 497–498.

33. For expansion on some of these points, see K. N. Lewis, *U.S. Strategic Force Planning: Restoring the Links between Strategy and Capabilities*, RAND P-6742 (Santa Monica, CA: Rand Corporation, 1982).

34. United States Department of Defense, *Annual Report to the Congress, Fiscal Year 1988* (Washington, DC: U.S. Government Printing Office, January 1987), 41.

35. Ibid., 20–41.

36. Murray and Viotti, *The Defense Policies of Nations*, 61.

37. Ibid., 74.

38. The U.S. force posture for the Cold War era is spelled out in the *Defense Department Report: FY 1980* (Washington, DC: U.S. Government Printing Office, 1979), 114, 125, 136, 140, 159, 178, 181, and 195. See also *The Military Balance: 1978–1979* (London: International Institute for Strategic Studies, 1978), 5–8.

39. Ibid., 70.

40. Murray and Viotti, *Defense Policies of Nations*, 74.

41. Ibid., 82–86.

42. Ibid., 73.

43. Ibid., 107–108.

44. Ibid., 109.

45. Ibid., 110–111; see also J. B. Stein, *From H-Bomb to Star Wars* (London: Lexington Books, 1984), 34–36.

46. B. Blechman and S. Kaplan, *Force without War: U.S. Armed Forces as a Political Instrument* (Washington, DC: Brookings Institution, 1978), 22.

47. Ibid., 38.

48. A. T. Mahan, *The Influence of Sea Power upon History, 1660–1783* (Annapolis, MD: U.S. Naval Institute, 1890).

49. J. H. Belote and W. M. Belote, *Titans of the Seas: The Development and Operations of Japanese and American Carrier Task Forces During World War II* (New York: Harper and Row, 1975), 301–305.

50. N. Friedman, *The U.S. Maritime Strategy* (London: Jane's Publishing, 1988), 102–103.

51. Ibid., 110.

Chapter 5

An Era of Transition: The United States in the Post–Cold War Era, 1989–1995

This chapter illustrates the linkage among the strategic situation in which the United States presently finds itself, its national security strategy and national military strategy, and the kinds of weapons and forces the United States may develop to support power projection and achieve those strategic objectives.

THE STRATEGIC SETTING

The last 30 years have witnessed a maturation in the global political situation of almost biblical proportions. The collapse of the Soviet state, the termination of the Cold War, and the emergence of the remnants of the Soviet empire as decolonized and independent states present unique opportunities for global security strategies. The United States has moved into an era of revolutionary international and domestic political change, which will require the conceptualization of fresh approaches if the nation is to retain its present position as a global superpower. Issues ranging from the environment, military space exploitation, rising Third World powers, and the continued threat of nuclear proliferation—and the subsequent issues of arms control—face U.S. policy makers, but with a changed structure and new dimensions.

During the twentieth century, the international political system has taken on a global character in which states and regional subsystems have become increasingly interdependent. By the next century, the boundaries of the current international system may even extend beyond this planet, and the use of space for military purposes will provide both a challenge and an opportunity for the United States. It is not clear at this point what the opportunities and the risks of space are, but certainly, the challenges are of a greater magnitude than those the U.S. faced as it encountered the intercontinental ballistic missile (ICBM) and territorial issues of the Cold War.[1]

Already, space has become no longer simply a place to explore—but one in which to work. Development of the Strategic Defense Initiative (SDI or "Star Wars") and militarization of space provide a basis for international competition that potentially could take a violent form. Current U.S. space policy is based on the free access to, and the peaceful exploration and development of, space.[2] A critical challenge for U.S. defense policy, as mentioned in Chapter 4, is to secure these goals. The difficulty now is that the number of potential actors in the development and exploration of space has more than doubled. The number of potential Third World participants that will soon be capable of initiating launches into this territorial extension is strategically alarming. Future strategists will have to determine how the United States will address this international concern.

Similar challenges in the strategic setting result from changes in the international political system. During the Cold War years, the major, if not only, actor in the international political system was the nation-state (i.e., the United States, the Soviet Union, Great Britain, Japan, etc). The nation-state alone best provided for the security and well-being of the people who organized themselves within it. While this institution has survived the Cold War, what faces the United States and its allies in this era of transition is the problem of unguided fledgling states.

Owing largely to decolonization, the United States is faced with a growing challenge as these new polities claim positions in the international system. For example, nations such as Yugoslavia, Romania, and the Baltic states of Estonia and Lithuania demand political and economic independence and the right to commerce, trade, and the wealth of the free world. They demand a voice in determining the direction of the international environment, and want to play a role in it. The issues and concerns of these nations require a U.S. security and military strategy focus. The failure to strategically address and prepare for the inherent difficulties of immature states may have notable ramifications as the proclaimed roots of international terrorism are certainly found in the unsolved reform issues of the Third World nations.[3]

The Cold War structure of the international political system was an intricate and multitiered web of power relationships dominated primarily by military matters. The bipolar, antagonistic relationship of the United States and Russia colored all relationships and was a factor in all international calculations. However, the Cold War is over, and the most significant international roles now developing seem to be economic.[4]

These emerging economic powers are the People's Republic of China,

Europe, and Japan. At the close of the Cold War, in late 1988, the private commission on Integrated Long Term Strategy pointed out: "By the year 2010, China and Japan will have the economic capacity to act as major world powers. Unless restructuring produces startling new gains, the Soviet Union's share of the world economy will shrink."[5]

China's role may be less predictable, but its role will surely be significant. As the remaining bastion of modern communism, the situation in China could theoretically develop in a number of ways, for example: it could grow stronger, as did the post-Soviet model of command economy and communist political arrangements; it could fail and splinter into a series of immature and concomitant nations, as did the Soviet Union; or it could continue its slow development as a semi-capitalist economy flirting with democratic reform.[6]

Similarly, the major alterations in the political and economic systems of Eastern Europe in 1989 and 1990 and the dissolution of the Warsaw Pact suggest a new role for Europe in the international system. Further, beyond these major powers are numerous, more diverse, and smaller centers of power, both regional and economic-based, which require U.S. strategic attention. Robert L. Pfaltzgraff, Jr., of the Institute for Foreign Policy Analysis claimed, "The economic, political and military power of nations like Korea and India will grow exponentially and will be relative to those in Western Europe."[7]

Japan will not remain simply an economic threat; it may also project a greater military role in the next decade.[8] In support of President George Bush's post–Cold War collective security strategy, U.S. Secretary of Defense Dick Cheney stated, "We expect Japan to move ahead with improvements in its own defense infrastructure, . . . improve its anti-invasion defenses and . . . defend its sea lines of communication."[9] However, it is not clear if Japan is willing to assume this expected new role; the provision to the United States of basing rights and economic compensation for participation in regional conflict may be the extent of its pledge. Author and president of Dietrich Industries, William Dietrich, claimed that the Japanese know that, "military superiority is no longer meaningful." No doubt, the Japanese have learned from experience. Dietrich added that to the Japanese, "the pursuit of power by military means is a dead-end street, but the economic gambit offers interesting possibilities."[10] In this new era, a strategy of shared defense in support of national interests is critical. The United States must be comfortable with the dedication of its allies as it decreases its force structure.

As a result, U.S. defense policy makers and the National Command Authority will have to deal with an increasingly more complex set of interrelationships, each of which has the potential to disrupt international stability and threaten U.S. national security interests. The challenge for U.S. national security and military strategy will be to determine methodologies, weapons, and forces that incorporate these complexities, and to present alternatives to conventional solutions.

Another particularly difficult problem in the current strategic setting is the continued danger of the proliferation of nuclear and high-tech conventional weapons, and what the Commission on Integrated Long Term Strategy called "the worldwide diffusion of advanced weapons." The commission report stated:

> The relationship between the major and minor powers will change by the early 21st century. Today the United States and Russia can often decisively influence the military postures of smaller states by making weaponry available or denying it. In the years ahead, weapons production will be much more widely diffused, and the superpowers . . . will have less control over transfers of advanced systems. Many lesser powers will have sizeable arsenals.[11]

As the commission report illustrates, despite whatever the United States does to control the proliferation of advanced weapons, there is no doubt about the increasing risk that unstable regimes, which potentially may be hostile to U.S. interests, will acquire these weapons.

The number of nations that have or will have nuclear weapons is increasing. Russian nuclear scientists may feel compelled to take a mercenary position because of their dire economic position, and may desperately market their expertise internationally. Inflamed by hard times, this reaction is not unlike the post–World War II U.S. and Soviet competition for Hitler's weapons designers. German scientists and engineers went east and west by the thousands, with many of them finding productive careers with their past enemies. For example, Verner Von Braun, former Nazi and designer of the V-2 rocket, was hired by the United States to become head of the group that put the first U.S. satellite into orbit. Von Braun eventually became one of the major architects of the National Aeronautical and Space Administration (NASA) launch systems. A number of nations are able to procure funding in support of this type of activity; it is estimated that by the year 2000, more than 40 nations, including some that are relatively poor and underdeveloped, will have the ability to build, deliver and detonate nuclear weapons.[12]

In January 1992, the London *Daily Telegraph* quoted U.S. Defense Secretary Dick Cheney as being "concerned that the Soviet nuclear arsenal will start to leak on to the international black market." The U.S. Defense Intelligence Agency was convinced that Pakistan was one of the nations that has gone beyond a theoretical nuclear capability and has developed "a nuclear device it could use."[13] Additionally, The *Philadelphia Inquirer* quoted Secretary of Defense Dick Cheney as saying, "The former republics of the Soviet Union are still building, deploying and selling nuclear weapons despite their commitment to reduce and control those arsenals, . . . which is unacceptable behavior in the long term."[14]

The challenge lies in how to control such proliferation. Conventional approaches developed during the Cold War era with agreements like the Nuclear Nonproliferation Treaty of 1968 or strategic deterrence as a nonproliferation strategy have failed to maintain the international status quo. Therefore, U.S. defense policy makers are confronted with the need to proliferation, which will require less-than-conventional methods.[15]

The remarkable evolution of mass media changed a national information resource into a global conglomerate that is capable of penetrating every nation of the world.[16] The influence of mass media on U.S. security and defense policy arises from changes that have taken place in the technology of the media and their worldwide diffusion. Information on events is available to policy makers in vast amounts and with an uncanny immediacy. In the Gulf War, Americans watched a high-tech conflict where weapons were delivered to their targets by pilots flying 500 miles per hour and watching what appeared to be a television. Never before was the surgical devastation of war portrayed so quickly, with so much emotion, and to so many people. William C. Adams, author of *Television Coverage of International Affairs*, said that the impact of media on U.S. policy makers is two-fold: "First, the mass media play active roles in U.S. policy development and execution. In specific situations, the media are active players in the policy making process—directly affecting specific policy outputs and events."[17] Adams also noted that: "Government recognizes the power of this global media and uses the media as a tool in international policy making."[18] "Second, mass media shapes the tone, style and emphasis of U.S. policy making in various ways and to varying degrees, both in specific instances and across the political spectrum."[19]

It is clear from Adams's analysis that mass media exerts a pervasive influence on the public and on U.S. policy makers.

Domestically, the United States is experiencing new challenges in the arenas of racial equality, economic unity and growth, illegal drug use, national debt, and environmental issues. Facing defense policy makers is a mix of turmoil similar to the events at the turn of the last century. The agenda for the resolution of these enigmas calls for unprecedented tactics and new, energetic security and defense strategies.

Racism is a growing problem in this post–Cold War era. In April 1991, the acquittal of four white police officers in Los Angeles, who were tried for excessive use of force while arresting a black man, escalated racial tensions, which resulted in riots, the loss of twenty-three innocent lives, and citywide arson and looting. The national media announcement of the verdict sparked similar reactions among minorities in other major cities across the United States. The magnitude of this response indicates the presence of underlying national tension. Author James Hunter asked, "What if these events are not just flashes of political madness but reveal the honest concerns of communities engaged in a deeply rooted cultural conflict?"[20] This problem, and its effect on the U.S. national interests, is linked to other equally critical and virulent domestic issues that impact U.S. national interests, is linked to other equally critical and virulent domestic issues that impact U.S. security interests—for example, illicit drug use, the trade imbalance, problems in education, and the social gap between rich and poor. The multifaceted model of racism in the United States has taken on epidemic proportions and transcends all levels of national, state, and local leadership as well as the whole of society. This epidemic, which has remained unsolved over years of attempted reform, is especially viral when seeded by other problems and provides a frightening example of insurrection in the United States. Left unchecked, its potential for violence, political imbalance, social unrest, and lethality is significant.

A strong U.S. economic strategy is a prerequisite for future domestic solvency as well as U.S. credibility in the international political and economic system. Labor, management, industrial growth, and U.S. corporate strength are all concerns affecting national interests. Amplifying this view, Robert Hormats, in *The Roots of American Power*, analyzed the ability of the United States to mobilize for the liberation of Kuwait in 1990. According to Hormats, "America's capacity to execute such a vast campaign was no accident."[21] Hormats suggested further that "success was made possible by an underlying economy able to turn out great volumes of extremely advanced military equipment."[22] The predictability and stability of American economics is disconcerting. While the United States is expected to maintain military preeminence in the coming decades, trends in U.S. economics, from Hormats's perspective, indicate that "domestic savings are at a historical low and that investment in education, industry, infrastructure and vanguard technologies is inadequate."[23]

Events of the past confirm that U.S. economic strength is tied to the achievement of national objectives. If the United States fails to maintain a strong macroeconomic performance, its ability to influence global economics and pursue a strategy that expands these market economies will be put in jeopardy. The relationship of U.S. economics and international power dictates that the nation guard against what Hormats called "imperial overstretch . . . and domestic underperformance."[24] The important point is this: a paradox exists between U.S. defense commitments and the U.S. economy, as an expanded forward presence throughout the world is not only a drain on U.S. economy, it is also vital to it. Consequently, policy makers and the National Command Authority will require innovative and imaginative solutions to support national security and military objectives and simultaneously approach and resolve job exportation, trade imbalance, and the decaying commitment of the U.S. corporate structure.

In addition, the National Security Council had described the international trade in illegal drugs as "a major threat to U.S. national security."[25] Domestic demand seems unrelenting, and the ability of the cartels and other organized production, transportation, and sales structures test the fabric of U.S. ideology and the nation's objective of individual and national progress. Few if any other nations have a drug problem proportionate to that of the United States. How the decision makers elect to employ national defense assets in an effort to eradicate this problem—either at the supply or the demand ends—remains to be seen. The current utilization of defense assets appears to be a dysfunctional, superficial application of national resource.

The aggregate debt of immature Third World nations was projected to reach $1.3 trillion in 1990, according to the World Bank.[26] The U.S. national debt is quickly reaching $5.5 billion, resulting in knee-jerk responses to defense spending authorization at the close of the Cold War era. This development follows historical trend lines and will undoubtedly effect the same consequences: a rapid drawdown and economizing of force growth and weapons development, leading to an uncontrolled surge and rapid rebuilding during times of conflict—at a cost of human life.

The United States traditionally has been a creditor nation and has used this status as a tool to influence international political circles. However, the current economic drift indicates a change to even this pattern. Nations consisting mainly of evolving Cold War allies, like Japan and Germany, look to be the economic leaders of the new world order. U.S. resolve to redirect economic projections and the inherent global power accompanying economic prosperity and the label of creditor government thus affect national security.[27]

Global environmental concerns over issues such as ozone depletion, global warming, food security, and deforestation, to name but a few, are directly affecting U.S. national security. Furthermore, global stress concerning these problems is contributing to international political friction. The roles of U.S. diplomacy and, potentially, defense forces may require new technologies and training to address environmental concerns. Recognizing that this application is other than conventional, defense planners may find that the nation considers these missions an appropriate use of defense assets in future competitive peacetime endeavors.

Is it possible to maintain adequate national security in this era of change and at the same time behave in such a way internationally as to induce potential enemies to join what appears to be a steady march toward peace? Can the United States retain a base force capability sufficient to deter, maintain conventional forces sufficient to meet local and regional aggression, and resist possible encroachments on U.S. interests and those of its allies? Part of the answer lies in intelligent arms management, including adopting a successful deterrent strategy in the Cold War and transferring that strategy to the post–Cold War era; part lies in the gradual transformation of conventional land, sea, and air forces into specially designed force mixes, capable of achieving U.S. national security objectives in their entirety. The strategic setting of post–Cold War America, and the world, requires a new vision to achieve U.S. national objectives. To do so, the United States must rethink completely its policy positions. In a 1991, Aspen, Colorado, speech, President George Bush articulated his vision for "a new world order." That vision, as outlined in national interests and objectives, is discussed in the following sections.

U.S. National Security Strategy (Post–Cold War)

As we stand at the threshold of a new strategic era, the security strategy of the United States must transform to engage the emerging geopolitical situations now facing the American people. Despite the dissolution of the Soviet Union and its inherent ideology, the gravest mistake the National Command Authority could make would be to preserve the medium of past security strategies. The collapse of the Soviet empire is now a strategic success, but it can become a strategic failure. The United States cannot lack solid analysis and forward thinking concerning the issue of new national security strategy made to support the president's "new world order."

Following is a dialogue on the new security strategy of the United States. It examines each of its strategic interests and goals, using the outline provided by the National Security Council,[28] which details U.S. strategic interests and objectives.[29]

Maintaining the United States as a free and independent nation and protecting its people, institutions, and values is the core interest of Americans (see Table 5.1). Clearly, the United States has valued security, stability, and prosperity since its founding and is certain to continue doing so. The pursuit of this fundamental interest has not changed since the inception of the constitution, but the specific interests we derive from it have. For instance, since World War II, the deterrence of nuclear attack was a major interest derived from the desire to maintain U.S. freedom and independence. Now that the threat of nuclear attack from the Soviet Union is essentially gone, however, much less emphasis is given to this objective.

For more than 45 years, the strategy of containment that evolved from the pursuit of this national interest served to check the expansion of Soviet communism and provided a focus for U.S. policy planners. Now,

Table 5.1
The First U.S. Security Interest and Its Supporting Objectives

INTEREST

(1) Ensure the survival of the United States as a free and independent nation, with its fundamental values intact and it institutions and peoples secure.

OBJECTIVES

a. Deter any aggression that could threaten the security of the United States and its allies and--should deterrence fail--repel or defeat military attack and end conflict on terms favorable to the United States, its interests and allies.

b. Effectively counter threats to the security of the United States and its citizens and interests short of armed conflict, including the threat of international terrorism.

c. Improve stability by pursuing equitable and verifiable arms control agreements, modernizing our strategic deterrent, developing systems capable of defending against limited ballistic-missile strikes, and enhancing appropriate conventional capabilities.

d. Promote dramatic change in the Soviet Union, while maintaining firm policies that discourage any temptation to new quests for military advantage.

e. Foster restraint in global military spending and discourage military adventurism.

f. Prevent the transfer of militarily critical technologies and resources to hostile countries or groups, especially the spread of chemical, biological and nuclear weapons and associated high-technology means of delivery.

g. Reduce the flow of illegal drugs into the United States by encouraging reduction in foreign production, combatting international traffickers and reducing demand at home.

Source: President George Bush, *National Security Strategy of the United States* (Washington, DC: U.S. Government Printing Office, August 1991).

however, the end of the Cold War and its accompanying circumstance of bipolarity finds containment gone from the national security strategy. The United States has awakened to new possibilities in global geopolitics that would have been unimaginable a few years ago. The objectives supporting the first national interest of ensured freedom and independence have changed from a focus on stopping Soviet expansion to one on responding to major regional threats. This focus is now a common thread throughout the new strategy articulated by the National Security Council.

Support of these objectives through diplomacy is a political watchword that should be used with caution, as goodwill without common sense can prove to be costly. The United States will continue to pursue strategic arms reduction treaties to reduce the threat of nuclear war; even if the Soviet Union is no longer the perceived threat, there are other nation-states capable of this effect. Arms control is an important element of strategy, and to foster restraint in military adventurism and spending, the United States must work toward the continued ratification of treaties, old and new, and the establishment of new alliances. The prevention of technology transfer to potentially hostile leaders may prove to be one of the more difficult objectives to master. As discussed in the strategic setting, "tech-mercenaries" are already emerging. How the U.S. plans to neutralize these nuclear and high-tech mercenaries will be a critical task for defense planners and will impact heavily on U.S. intelligence services and budget.

The second national interest, of maintaining a healthy U.S. economy at home and abroad, is derived from two beliefs: first, that unrestricted free trade will do for the international system what laissez-faire economics did for the nation, and second, that the U.S. standard of living depends on the ability to secure certain raw materials, invest overseas, and have access to foreign markets.

The objectives supporting this economic interest confirm America's longstanding commitment to maintaining international agreements to help keep the high seas open to free and unrestricted navigation (see Table 5.2). Overall, the United States accounts for 22 to 25 percent of the major forms of global economic activity, and has done so for twenty years or more.[30]

The National Security Council sets the objectives under this second national interest because it perceives five major economic challenges to the United States. First, by promoting democratic values of political choice, human rights, and self-determination, the United States can foster global hope and recovery, which in turn will bring about increased prosperity, free trade, and an interlinked, international economic system.[31]

Second, the United States must take the lead in breaking down nationalist barriers that prove counterproductive to world economics by promoting market principles. Working through institutions like the World Bank, the International Monetary Fund, and the General Agreements on Tariffs and Trade, the United States can break down these barriers.[32]

Third, security and stability are the foundation of economic growth, in the United States and internationally. Goodwill and assured access to needed resources can be achieved through a policy of assistance that strengthens the bonds of collective defense. Simply put, if the United States and its allies invest in a shared capability of defense, that security

Table 5.2
The Second U.S. Security Interest and Its Supporting Objectives

INTEREST
(2) Maintain a healthy and growing U.S. economy to ensure opportunity for individual prosperity and resources for national endeavors at home and abroad.
OBJECTIVES
a. Promote a strong, prosperous and competitive U.S. economy.
b. Ensure access to foreign markets, energy, mineral resources, the oceans and space.
c. Promote an open and expanding international economic system, based on market principals, with minimal distortions to trade and investment, stable currencies, and broadly respected rules for managing and resolving economic disputes.
d. Achieve cooperative international solutions to key environmental challenges, assuring the sustainability and environmental security of the planet as well as growth and opportunity for all.

Source: President George Bush, *National Security Strategy of the United States* (Washington, DC: U.S. Government Printing Office, August 1991).

and stability will enhance the economic prosperity of all actors involved.[33]

Fourth, the world is afflicted with any number of transnational threats, including drugs, AIDS, terrorism, and environmental damage, to name but a few. U.S. involvement in the eradication of these deterrents to world health will in turn foster conditions suitable to global well-being.[34]

Fifth, the United States is, and always will be, the principal actor in meeting global human needs. When disaster strikes, whether natural or otherwise, efforts taken to reduce human suffering quickly and tangibly will foster global well-being and promote an environment suitable to economic growth and exchange.[35]

The third U.S. national interest, of maintaining a healthy, cooperative, and politically vigorous relationship with allies and friendly nations, closely parallels the first and second interests. This seems to be a thread designed to tie together U.S. foreign policy in support of mutual cooperation on the issues of defense, human rights, world health, economic prosperity, and world stability. Coalitions constructed of free and amicable nations will secure an environment of peace for future generations, and U.S. diplomacy is to be a catalyst in this endeavor.[36]

Table 5.3
The Third U.S. Security Interest and Its Supporting Objectives

INTEREST

(3) Maintain a healthy, cooperative and politically vigorous relations with allies and friendly nations.

OBJECTIVES

a. Strengthen and enlarge the commonwealth of free nations that share a commitment to democracy and individual rights.

b. Establish a more balanced partnership with our allies and a greater sharing of global leadership and responsibilities.

c. Strengthen international institutions like the United Nations to make them more effective in promoting peace, world order and political, economic and social progress.

d. Support Western Europe historic march toward greater economic and political unity, including a European security identity within the Atlantic Alliance, and nurture a closer relationship between the United States and the european Community.

e. Work with our North Atlantic allies to help develop the processes of the Conference on Security and Cooperation in Europe to bring about reconciliation, security and democracy in a Europe whole and free.

Source: President George Bush, *National Security Strategy of the United States* (Washington, DC: U.S. Government Printing Office, August 1991).

The objectives supporting this interest will require a sustained effort in promoting new alliances and rekindling old ones (see Table 5.3). Every positive aspect of world peace is linked to the strengthening of international institutions like the United Nations or the North Atlantic Treaty Organization. The support of these institutions and alliances ensures economic growth, political stability, and world order. The United States must stay inexorably bonded to a philosophy of reconciliation, cooperation, mutual interest, and partnership via alliance and treaty. U.S. interest will always set the pace in world affairs, a topic on which American interests must remain focused.

The fourth national interest, maintaining a stable and secure world, where political and economic freedom, human rights and democratic institutions flourish, reflects the U.S. dedication to world peace. In this new era of diminished threat, the National Security Council knows that a stable and secure world is a prerequisite if the United States is to achieve its national objectives in all spheres. A strategy that expands and strengthens political and economic freedom, emphasizing human rights

Table 5.4
The First U.S. Security Interest and Its Supporting Defense Policy

INTEREST

(1) Ensure the survival of the United States as a free and independent nation, with its fundamental values intact and it institutions and peoples secure.

DEFENSE POLICY

a. Maintain a force that will serve as a credible deterrent--capable of providing a reliable early warning, and consisting of modern nuclear forces that can support a spectrum of response options.

b. Maintain a defensive system for global protection against limited strikes (GPALS).

c. Maintain a mix of survivable and capable nuclear offensive and defensive forces.

d. Maintain a high quality force.

e. Sustain intelligence capabilities.

f. Promote non-proliferation and technology security.

Source: U.S. Secretary of Defense Dick Cheney, *Annual Report of the Secretary of Defense to the President and the Congress* (Washington, DC: U.S. Government Printing Office, 1991).

and the establishment of democratic institutions, will transform the ways and means to that end.

The objectives outlined to support this interest are designed to limit regional turmoil through a close monitoring of the military balance of power (see Table 5.4). To keep this objective on track, the United States will rely on its intelligence collection capability to detect problems and its diplomatic assets to resolve them. Again, human rights, social progress, and economic growth are paramount, and the systems inherent to the U.S. policy must be finely tuned to monitor these issues. Low-intensity conflict will surface in the presence of nonreform on issues important to a nation's populace. Many Third World states, like Peru in Latin America and Cambodia in Southeast Asia, repeatedly fail to address the issues of agrarian and economic reform. A failure to resolve these issues invariably brings insurrection and terrorism onto the scene, shaking the foundations of global stability. U.S. economic assistance programs and international exchange programs under the Foreign Assistance Act help to quell unrest and eventually secure economic progress

Table 5.5
The Second U.S. Security Interest and Its Supporting Defense Policy

INTEREST

(2) Maintain a healthy and growing U.S. economy to ensure opportunity for individual prosperity and resources for national endeavors at home and abroad.

DEFENSE POLICY

a. Maintain a peacetime forward presence--at reduced levels--for crisis response and escalation control.

b. Maintain and tailor a force into three categories:

1. Forces for the Pacific hemisphere
2. Forces for the Atlantic hemisphere
3. Forces for contingencies

c. Continue investment in research of new technologies

Source: U.S. Secretary of Defense Dick Cheney, *Annual Report of the Secretary of Defense to the President and the Congress* (Washington, DC: U.S. Government Printing Office, 1991).

and social political progress. In part, this will diminish the effects on U.S. security, for example, by stemming the tide of immigrants and refugees, and by diminishing the drug trade at the cultivation level.

Certainly, the U.S. has not reached a stage where there are no more hazards to its security. The world is still a dangerous place, and America must remain prepared to counter a variety of threats. Regional and global security will remain important to the security of the United States, and even though the nation is economizing its force structure, it has also developed a new defense strategy to serve as the conceptual framework for its future forces and weapons.

U.S. Military Strategy (Post–Cold War)

As the Cold War era closes, the United States is witnessing changes in Eastern Europe that reflect the success of the enduring U.S. national strategy (see Table 5.5). Improvements in weapons and forces and the strength of its European alliances helped to secure this end. It is disconcerting that neither the United States nor its NATO allies have the ability to guarantee that the changes in Russia will continue along a democratic path.

Despite a positive manifestation of reform toward liberal democracy, the United States must remain poised to augment its defense effort. As the nation economizes its global force structures, it must be cautious not to degrade the perishable endowment of its leadership, technology, and industrial defense base. To do so would render the United States incapable of executing a quick recovery toward warfare, should the need arise.

The Russian Command Authority is reducing significantly, but not abolishing, its war machine. The Russians will not diffuse their strategic nuclear capability, for that remains their only link to superpower status. As the diffused Soviet war machine transforms itself into a new framework under another appellation, the potential for conflict genesis and escalation to global war will not vanish. If the negotiations on conventional forces in Europe (CFE) fail (in the wake of the collapse of the former Soviet Union), the ability of NATO forces to predict conflict and provide early warning will be critical. As such, NATO forces must now develop the ability to provide for security beyond that established by CFE.

Because global ballistic missile proliferation is a possibility—as is accidental launch—a major defense policy is the provision of early warning, with a capability to respond. Global Protection against Limited Strikes (GPALS) provides for this eventuality. Relying on modified Strategic Defense Initiative (SDI) technology, this system will protect the U.S. mainland against limited strikes. Furthermore, U.S. forces would be protected, as would its allies. This system is smaller than the original SDI design concept and less expensive, it is expected to follow the original deployment schedule of SDI.

The defense policy must assume proliferation in the Third World of nuclear weapons and long-range missiles, as well as chemical, and biological weapons of mass destruction. The changing geopolitical environment and subsequent defense strategy requirements dictate new roles and mission for U.S. forces. These new roles and missions impose requirements for forces that are smaller but heavily mobile, highly trained, and well-equipped with both lethal and nonlethal capabilities. They must be solid projectors of power, utilizing improved delivery platforms and sustainment systems.

The maintenance of a smaller, high-quality force in lieu of the large forces of the Cold War era is a new defense policy objective. Demands for a higher-quality force require force developers to concentrate on recruitment, retention, training, education, and more important, compensation that at least parallels that of civilian counterparts.

A more responsive intelligence collection and analysis capability is needed in the post–Cold War era. While the United States certainly has the most highly developed and sophisticated system available in the

Table 5.6
The Third U.S. Security Interest and Its Supporting Defense Policy

INTEREST
(3) Maintain a healthy, cooperative and politically vigorous relations with allies and friendly nations.
DEFENSE POLICY
a. Maintain existing alliances and other security partnerships.
b. Engage in arms control as part of a preplanned coordinated effort to enhance global security.
c. Share the responsibility for the common defense with our allies.

Source: U.S. Secretary of Defense Dick Cheney, *Annual Report of the Secretary of Defense to the President and the Congress* (Washington, DC: U.S. Government Printing Office, 1991).

world, its demonstrated ability to shift target priorities, conduct cross-region analysis, and integrate all available systems into a collective effort needs improvement. A case for improvement is discussed at the conclusion of this chapter.

The United States and its allies must plan for the continued protection of existing sea lanes of commerce through the employment of conventional maritime strategy and long-range air power (see Table 5.6). The U.S. ability to engage in out-of-area conflicts and crises must be not only maintained but improved. As new, immature nation-states emerge from the process of decolonization, the inherent security challenges will increase in intensity and sophistication.

The continued superiority of U.S. maritime forces is critical if defense policy is to include maintenance of the capability to project power globally as part of a flexible response strategy. U.S. warships will modernize for increased lethal capacity, theater endurance, and response time. As such, current trends toward diminishing overseas basing rights must be halted and quickly reexplored. Diplomatic efforts in support of defense linkages must effect change in this arena—or U.S. defense plans will suffer significantly.

The defense-basing issues are critical to the defense policy of tailoring a force into three categories. These new force structures include the use of forward-based and forward-deployed forces supporting the second security interest of the United States. Collectively, these forces provide the appropriate level of response capability and deterrence on a global scale. The forces for the Pacific hemisphere consist of forward-based and forward-deployed units with naval-, ground-, and air-reinforcing forces remaining in the United States. The forces for the Atlantic hemisphere consist of forward-based and -deployed forces focusing on the European theater. This force structure also consists of heavy forces dedicated to Southwest Asia, and it is augmented with forces for contingencies based in the United States. The forces for contingencies are ground, naval, and air forces based in the United States—they are capable of worldwide deployments on short notice but rely heavily on reserve component augmentation for support.

The loss of much of the vital defense industry base during the recent economization of force requires the industrial leadership to rethink and reformulate new strategic defense roles and acquisition policies. In the event a global threat emerges, the defense strategy of the 1990s may require the rapid procurement of high-tech weapons systems. However, it is unlikely that the United States will face a threat of this nature in the near term. Perhaps the future will prove differently, but the current strategic setting does not suggest a trend toward a global threat. In the coming decade, even greater importance will attach to our ability to develop and exploit advanced technologies for many applications, including new, technologically advanced weapons systems, satellites, and information processing (C3I) systems.[37]

Concepts of mutual defense and shared responsibility—in view of our economization of forces—now form an even more critical part of defense policy. Alliances have always helped to promote world security and regional stability. The United States presently maintains seven official alliances (see Table 5.7). There are also less formal agreements, such as bilateral relationships, that the United States keeps in order to foster international dialogue and the improved cooperation and integration of coalition military forces.

Arms control agreements—part of the alliance packages—help to make international relationships a bit more predictable. Nuclear, Biological and Chemical agreements have always been critical policy issues, now space and its potential for nuclear weapons developments is also a concern. This is an issue which followed the U.S. out of the Cold War—given technological advances it is now more than just a perceived threat. Several dialogues on this issue are ongoing, for example, the Defense Space Talks (DST).

Shared responsibility for defense requires that the interests of the United States and its allies remain closely tied. As the United States pairs down its forces, this has become a paramount issue. Sharing the cost and responsibility for contingencies and for the conduct of long-term planning for mobilization and sustainment are all part of this requirement. For example, the United States has not only requested that Japan and

Table 5.7
The Concept of Mutual Defense and Shared Responsibility

*The North Atlantic Treaty Organization (NATO) Alliance
*The Australia-New Zealand-United States (ANZUS) Alliance (although U.S. obligations to New Zealand are suspended as a result of New Zealand's decision to ban U.S. nuclear-powered and nuclear-capable ships from its ports)
*The Treaty of Mutual Cooperation and Security between the United States and Japan
*The Mutual Defense Treaty between the Unites States and Korea
*The Mutual Defense Treaty between the United States and the Republic of the Philippines
*The Southeast Asia Collective Defense Treaty (which remains in effect on a bilateral basis with Thailand)
*The Inter-American Treaty of Reciprocal Assistance (the RIO Treaty)

Source: U.S. Secretary of Defense Dick Cheney, *Annual Report of the Secretary of Defense to the President and the Congress* (Washington, DC: U.S. Government Printing Office, 1991), 8.

Korea assume increased monetary responsibility for the defense of their regions, but that they also begin the increased development of the necessary forces—air, ground, and sea—to provide for their security.

Developing effective responses to regional conflicts affecting U.S. interests translates in part to developing forces capable of responding to destabilizing factors wherever they may be (Table 5.8). While large conventional forces were, in essence, perfect for their era, the new world order dictates response capabilities that will project the necessary force into affected areas in order to respond to unrest caused by a weak or embryonic government. Low-intensity conflict differs from conventional conflict and requires more innovation than deterrence can offer. For example, special operations forces—under the umbrella of the contingency force—are designed to assist in this countervailing defense policy.

Support for counterdrug operations is an increasingly visible and important defense policy. There are in-place forces supporting these operations currently, but their productivity is marginal. Issues regarding posse comitatus, in-country military operations with the support of host nation governments, and so forth will need to be resolved before this defense policy can be effective.

Table 5.8
A New Strategic Era—A New Strategic Interest

INTEREST
(4) Maintain a stable and secure world, where political and economic freedom, human rights and democratic institutions flourish.
DEFENSE POLICY
a. Develop effective responses to regional conflicts affecting U.S. interests via innovative strategies that support representative government, integrate security assistance and promote economic development.
b. Use contingency forces with augmentation to counteract violence and promote nation building.
c. Detect and counter the production of illegal drugs, trafficking and use.

Source: U.S. Secretary of Defense Dick Cheney, *Annual Report of the Secretary of Defense to the President and the Congress* (Washington, DC: U.S. Government Printing Office, 1991).

WEAPONS AND FORCES

Responding to the realities of this new strategic era, U.S. Secretary of Defense Dick Cheney directed the construction of a package of weapons and forces suitable to the needs of U.S. national strategy, specifically, to deter aggression, react to regional crisis, maintain a forward presence, and rebuild a global war-fighting capability.[38] This force package (shown in Figure 5.1), which is called the "Base Force," is designed to meet U.S. strategic needs from peace to multiple regional crises.[39]

In peacetime, the responsibility for forward presence rests with the forces that are not focused on the objectives of strategic deterrence. U.S. commitments to allies via formal security agreements and treaties, as mentioned in Table 5.7, determines what portions of available peacetime forces are dedicated to the various regions of the world. Forces engaged in forward-presence missions assume and prepare for crisis response as part of their core capabilities. Crisis response forces serve in their forward-presence roles by participating in joint and combined training exercises in their respective regions of the world. This mix of forces provides the United States with projected power capable of extended deterrence in support of national interests. The forward presence of U.S. forces is an element of national power influencing world events and encouraging continued democratic and economic development as well as the veneration of human rights.[40]

Figure 5.1
The New Base Force Composition

FORCE COMPOSITION

		FY 92	BASE FORCE
STRATEGIC	Bombers	B-51 * B-1	B-52H * B-1 * B-2
	Missiles	1000	550
	SSBNs	34	18
ARMY	Active	16 Divisions	12 Divisions
	Reserve	10 Divisions	6 Divisions
	Cadre		2 Divisions
NAVY	Ships	530 (15 CVBGs)	450 (12 CVBGs)
	Active	13 Air Wings	11 Air Wings
	Reserve	2 Air Wings	2 Air Wings
USMC	Active	3 MEFs	3 MEFs
	Reserve	1 Division / Wing	1 Division / Wing
AIR FORCE	Active	22 FWEs	15 FWEs
	Reserve	12 FWEs	11 FWEs

CVSG: Carrier Battle Groups MEF: Marine Expeditionary Forces FWE: Fighter Wing Equivalent

Source: U.S. Government.

Figure 5.2 shows the national combat power increasing as the spectrum of conventional conflict escalates toward global war. This influx of power is made possible via the concept of using regional commanders-in-chief to tailor joint task forces and quickly react to the national threat. To react quickly, these task forces require pre-positioned war materials. Due to post–Cold War limitations in force sustainment and overseas basing, these war materials may not be readily available; therefore, maritime prepositioning is required to support national strategy. As some crisis-response forces are committed to one region, others remain dedicated to support of alliance commitments elsewhere. Simply put, the United States will not commit 100 percent of its base combat force to any one crisis.[41]

Figure 5.1 portrays the composition of the base force. The base force maximizes the capabilities of each component by integrating the active

Figure 5.2
The Concept of Base Force Support to the Operational Continuum

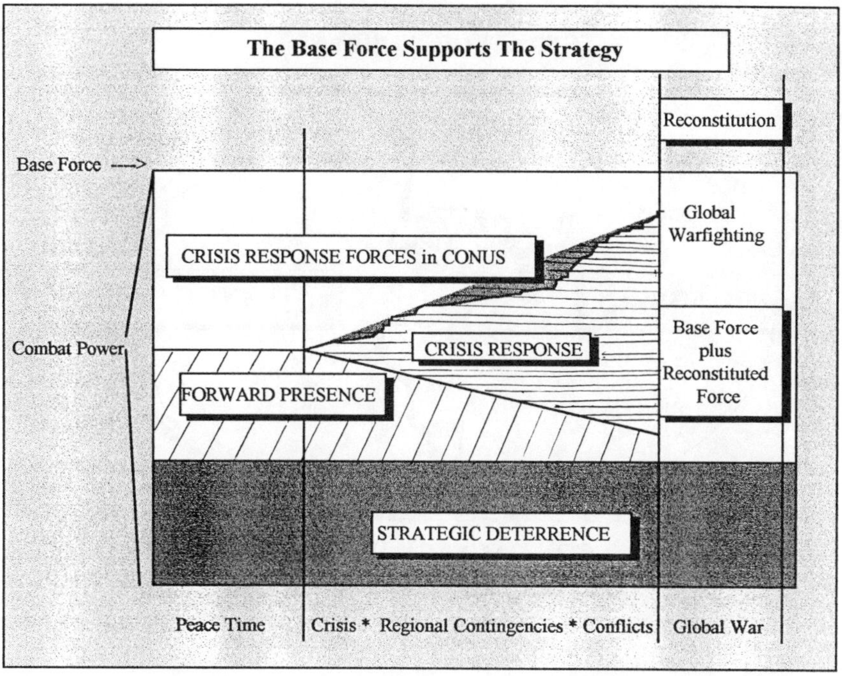

Source: U.S. Government.

and reserve forces of each service into an effective crisis-response force. This force is founded on the capabilities of Cold War adversaries, and on current and potential future threats. The base force concept meets the national security objective of increased responsibility and shared security by acknowledging the much-needed support to the United States by allied and friendly nations.[42]

The base force concept relies heavily on the use of reserve forces. Initially, the force structure focuses on active components to respond to regional crises; however, as the intensity of conflict moves on the spectrum toward global war, reserve forces are systematically integrated to increase combat power. The ability to provide timely and responsive reserve-force augmentation to the active base force requires the support of national will, the defense industrial base, and force mobilization structures. Therefore, the base force is designed to support, among others, the national security objectives of maintaining a peacetime forward presence—at reduced levels—for crisis response and escalation control. Furthermore, the base force concept promotes a strong, prosperous U.S.

Figure 5.3
Categories of Base Forces

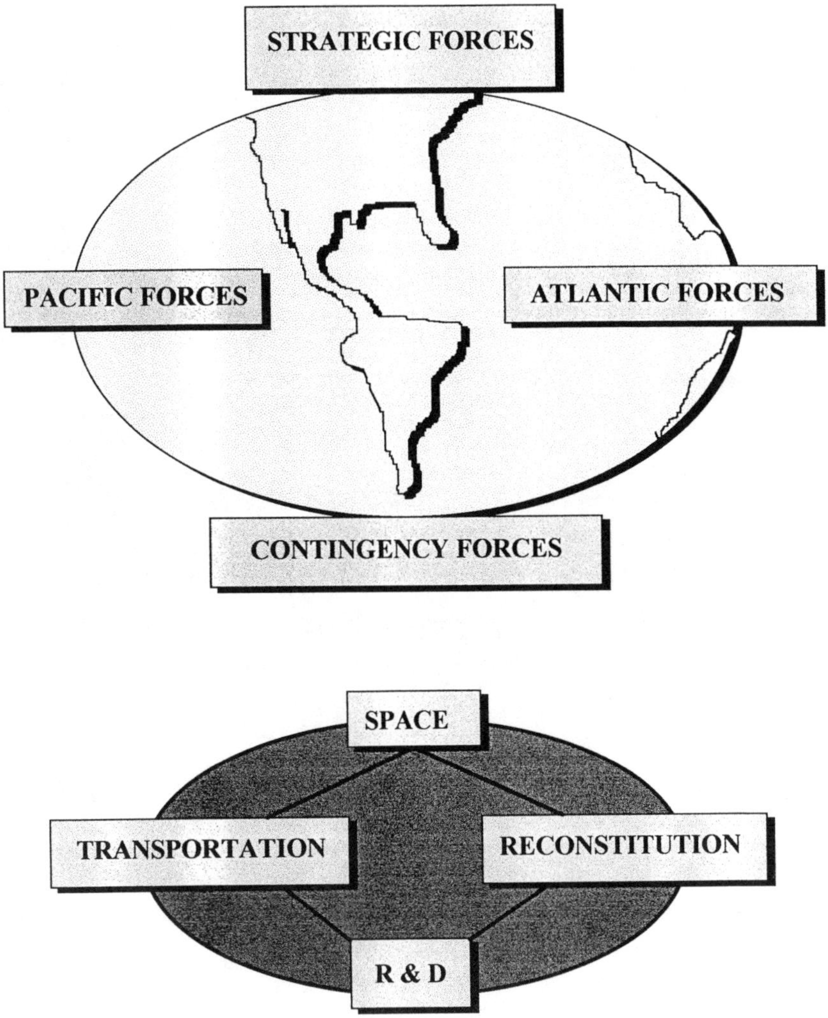

Source: U.S. Government.

economy by requiring the maintenance of a responsive, prepared defense-industrial base.[43]

Figure 5.3 illustrates the four base force packages—strategic, Atlantic, Pacific, and contingency—and the supporting capabilities of space, transportation, research and development, and reconstitution.[44]

Uncertainty in the status of the former Soviet nuclear arsenal, Third

World proliferation of mass destruction weapons, and global arms sales mandate the U.S. national security strategy requirement for an ability to deter and respond to a nuclear threat. The new strategic component of the base force supports this security objective by establishing a triad of modern, flexible, and survivable systems.[45]

One leg of the triad, Global Protection Against Limited Strikes (GPALS), is a planned space-based system capable of disrupting and destroying ballistic missile strikes launched on deployed U.S. forces, as well as on friends and allies of the United States. Another leg of the triad is the strategic air forces, consisting of long-range bombers, like the B-52, B-1, and the B-2, that maintain a nuclear and conventional capability matched only by the Soviets. The third leg of the triad consists of 500 "Minuteman" III and 50 "Peacekeeper" land-based intercontinental ballistic missiles (ICBMs), and 18 ballistic missile submarines.[46]

The Atlantic forces protect U.S. interests in Europe, the Middle East, the Mediterranean, Africa, and Southwest Asia. Forward-presence requirements in Europe are reduced but still existent because European security and stability remain inexorably tied to U.S. national interests. Other forces in the region include forward-stationed forces and rotational forces capable of flexing in size to meet the threat.[47]

Ground forces serving a forward-presence mission are arranged as a two-division Army corps. This corps is the foundation of a U.S.-led multinational corps, and it has also provided one division to a German-led multinational corps. These ground forces are specifically tailored to meet the new strategic needs of the United States and its allies. The force can conduct combat operations in Europe or project power anywhere in the region—reinforced by U.S. based units when required.[48]

The Atlantic air forces consist of three to four fighter wings, which can meet a wide range of theater tasks, and most important, the rapid achievement of theater air superiority. Additional air force tasks conducted in support of national interests are the suppression of theater enemy defenses, as well as strategic and tactical air strikes. Continental U.S.–based strategic bomber forces provide the long-range support arm and round-out the regional air forces.[49]

Naval and amphibious forces deployed to the Atlantic provide forward-presence and response capabilities throughout the seas of the region. These directly support U.S. interests of ensuring open sea lanes and ports and free access to the regional waters. U.S. Naval and Marine forces execute forward presence through their ability to rapidly project power via the sea, merging with Pacific naval forces to place two carrier battle groups and amphibious ready groups in theater. The capability to deliver marine ground forces, sustain combat operations, and conduct air–ground operations with carrier-based aircraft is critical to the protection of national interests in the region.[50]

In the Pacific region, the force is smaller than past structures. The region is vastly expansive, and it is economically important to the United States. The character of the region lends itself to a maritime base force comprised of half the total carrier and amphibious forces available to the United States, with a forward-deployed carrier battle group and one marine expeditionary force. The loss of basing in the Philippines forced one carrier battle group and one amphibious group to be home-ported in Japan.[51]

Army and air forces collectively provide support to the region as part of the total base force package. However, as U.S. emphasis on shared responsibility for regional security matures and the army and air force of South Korea develop, U.S. ground and air presence can be reduced.[52]

Regional crisis response forces include Alaska, Hawaii, and CONUS-based forces, which include a reinforced division, one fighter wing, and five carrier battle groups.[53]

Contingency forces serving as part of the base force structure consist of forces that can be tailored by the supported CINCs to support threatened national interests. Specifically, the ground force element contains an airborne division capable of air dropping or air landing combat elements to secure theater objectives. Another element of the ground force is an air assault division—with unique helicopter capabilities—and two heavy divisions, which are capable of rapid transport in support of crisis response.[54]

Contingency air forces are multiaircraft and tailored for rapid power projection to support ground forces and conduct specified air campaigns. The entire range of fighter, fighter-bomber, and long-range conventional bombers is provided by the air force.[55]

The navy and amphibious forces project regional power via carrier-based aircraft in direct support of amphibious forces or as part of a joint air campaign with U.S. and allied air forces. Attack submarines and surface-warfare ships support amphibious combat power and assist the United States in maintaining the unique capability of securing vital interests along foreign shores and up to, and even exceeding, 50 miles inland.[56]

The four supporting capabilities of transportation, space, reconstitution, and research and development (R&D) are each critical components of the base force concept. Ground, air, and naval forces directly support the military strategy, but the support structure is an intricate chain that cannot develop a weak link, as to do so would disrupt or even halt a deployment.[57]

Transportation assets in the form of strategic airlift, via the C-5 "Galaxy" and C-141 "Starlifters," as well as the new C-17, are required to ensure that the base force has a strategic mobility capability to rapidly deploy and sustain U.S. combat power in any region where national interests are threatened. Augmenting the strategic airlift is tactical airlift asset for intertheater use consisting of C-130 "Hercules" aircraft.[58]

Space forces are required to meet national security needs such as early warning; command, control and communications; and surveillance. Essential strategic defense, growing tactical demands, and the monitoring of treaties and alliances require effective spacecraft and infrastructure. The Secretary of Defense Dick Cheney specified that U.S. space forces should accomplish four tasks: space control—combat against enemy forces in space and their infrastructure; force application—combat against enemy land, sea, air, and missile forces; force enhancement—providing support to air, sea, and land forces; and space support—consisting of satellite control and launch capability. The importance of this new dimension of security is amplified by author James F. Dunnigan's comments in *How to Make War*: "Currently about 1,500 satellites are in orbit, 43 percent of them are Russian and 38 percent are U.S., the remaining 19 percent belong to western nations."[59] As space becomes an extension of U.S. security responsibility, the ability to employ space forces in support of U.S. armed forces becomes increasingly critical. To meet the challenges of this growing threat, a high priority is given to the development of space capabilities for early warning, deterrence, and direct support of conventional forces and tactical operations.

CONCLUSIONS

In the present world, humankind is up against certain problems that are now too big to be evaded. The problems of changing geopolitics, economics, and mass media all affect the decisions of world leadership and transform a nation's response to strategic needs. Compounding these problems is the continued need to survive as a people and to provide for the basic security of the nation.

The National Security Council has clearly recognized that the status of world affairs should not lead the United States to false notions of security.[60] Furthermore, American conservatism should not be responsible for a reluctance to recognize change and adapt. Therefore, the nation has developed forces and weapons that continue to provide for its national security strategy and interests. Will the base force be enough to accomplish these tasks in the future? Can America, politically, economically, and domestically, survive a protracted war? Even with a thinning force structure, can the defense department and the president adequately justify to the American people the billions spent annually on defense? Argument can be made that there is no direct threat to the security of the United States—at least in the near future. Therefore, under what pretense does the National Security Council justify a budget of hundreds of billions of dollars; and what can forward thinkers do to help?

Chapters 3, 4, and 5 have been dedicated to illustrating the linkage among a nation's strategic setting, national security strategy, military strategy, and the corresponding weapons and forces developed and packaged to support those strategies.

Changes in the international security system and domestic budget constraints mandate that the United States seek alternatives to conventional methods of forward presence and long-reach capabilities. The new renaissance of military technology is underway, and the military strategy of the future will have to explore, develop, and perfect new weapons and doctrine that extends beyond conventional boundaries.

Fielding systems that are low cost, effective, and most of all, innovative to the point that they are not yet understood or employed by potential adversaries may command popular support and create much-needed military leverage in the post–Cold War strategic environment. As the secretary of defense stated in a recent Defense Planning Guidance document for fiscal year (FY) 1992–1997, it is imperative to "train the officer corps to develop new concepts with the support of new weapons systems."

Nonlethal technology can supply weapons and methods that augment and, in some cases, replace conventional lethal technologies. The new technologies are designed to enhance deterrence by making the success of enemy attacks unlikely through destroying or degrading of the latters war-making machine. Weapons systems and technologies with this capability are likely to be fully supported by the American people. Several examples are in order, and that is the function of the next chapter.

NOTES

1. S. J. Hadley, "The Nature of SDI," expanded in the Johns Hopkins University Foreign Policy Institute (FPI) policy brief, *Thinking about SDI* (Washington, DC: Johns Hopkins Foreign Policy Institute, 1986), 16–17.

2. President George Bush, *National Security Strategy of the United States* (Washington, DC: U.S. Government Printing Office, August 1991), 22–23; see also Michael Vlahos, *The American Ethos and National Security* (London: Westview Press, 1990), 34–35.

3. H. J. Vetter and G. R. Perlstein, eds., *Perspectives on Terrorism* (Pacific Grove, CA: Brooks/Cole Publishing, 1991), 190–191.

4. G. C. Hufbauer and K. A. Elliot, "The International Economy and National Security: Where Do They Meet?" in *Global Security: A Review of Strategic and Economic Issues*, ed. B. M. Blechman and E. N. Luttwak (London: Westview Press, 1987), 91–92.

5. The Commission on Integrated Long Term Strategy, *Discriminate Deterrence* (Washington, DC: U.S. Government Printing Office, January 1988), 7.

6. W. R. Heaton, Jr., "The Defense Policy of the People's Republic Of China," in *The Defense Policies of Nations*, ed. D. J. Murray and P. R. Viotti (Baltimore, MD: Johns Hopkins University Press, 1989), 419–421.

7. R. L. Pfaltzgraff, Jr., *National Security: Ethics, Strategy, and Politics*, Foreign Policy Report (International Defense Publishers, February 1986; reprint, New York: Pergamon, Brassey's, 1988), 26. See also F. Calogero's article entitled, "Critical Remarks on European Security," in *Rethinking European Security*, ed. F. Cerutti and R. Ragionieri (Washington, DC: Crane Russak, 1992), 94.

8. G. Friedman and M. Lebard, *The Coming War with Japan* (New York: St Martin's Press, 1988), 63.

9. U.S. Secretary of Defense Dick Cheney, *Annual Report of the Secretary of Defense to the President and the Congress* (Washington, DC: U.S. Government Printing Office, 1991), 9.

10. W. S. Dietrich, *In The Shadow Of The Rising Sun—The Political Roots of American Economic Decline* (New York: Pennsylvania State University Press, 1991), 262.

11. Commission on Integrated Long Term Strategy, *Discriminate Deterrence*, 8.

12. Ibid., 9.

13. Armed Forces Information Service (AFIS/OASD-PA), Wire News, 8 January 1992, 16.

14. AFIS/OASD-PA, December 1991, 1.

15. P. S. Kronenberg, *Planning U.S. Security: Defense Policy in the Eighties*, study developed at the National Defense University (1982; reprint, New York: Pergamon Press, 1987), 93–95.

16. P. O'Heffernan, *Mass Media and American Foreign Politics: Insider Perspectives on Global Journalism and the Foreign Policy Process* (Norwood, NJ: Ablex Publishing, 1992), 85.

17. W. C. Adams, *Television Coverage of International Affairs* (Norwood, NJ: Ablex Publishers, 1987), 68.

18. Ibid., 87.

19. Ibid., 85.

20. J. D. Hunter, *Culture Wars: The Struggle to Define America* (New York: Basic Books, 1990), 33.

21. R. D. Hormats, "The Roots of American Power," *Foreign Affairs* 149 (Summer 1991): 132.

22. Ibid.

23. Ibid.

24. Ibid., 137.

25. Bush, *National Security Strategy of the United States*, 17.

26. Ibid., 20.

27. G. C. Hufbauer and K. A. Elliott, "The International Economy with a National Security Perspective," in *Global Security: A Review of Strategic and Economic Issues*, ed. Barry M. Blechman and Edward N. Luttwak (London: Westview Press, 1987), 93–94.

28. Bush, *National Security Strategy of the United States*, 3–4.

29. Ibid., 5–23.

30. General Colin L. Powell, Chairman of the Joint Chiefs of Staff, *The National Military Strategy 1992* (Washington, DC: U.S. Government Printing Office, 1992), 2–5.

31. Ibid., 5.

32. Ibid., 4–5.

33. Ibid., 2, 8.

34. Ibid., 7, 14–15.

35. Ibid., 2–4, 6.

36. Ibid., 3–4.

37. See the U.S. Secretary of Defense, *U.S. Defense Planning Guidance, Fiscal Year 1992–1997*, Secretary of Defense Control Number: x52264, Classified (SECRET/NOFORN/NO CONTRACT); the text referenced here is not classified.

38. Powell, *National Military Strategy*, 17.

39. Ibid.

40. Ibid., 17–18.

41. Ibid., 18.

42. Ibid., 19.

43. Ibid., 19–20.

44. Ibid., 20.

45. Ibid., 19.

46. Ibid., 20.

47. Ibid.

48. Ibid., 19–20.

49. Ibid., 21.

50. Ibid., 21–22.

51. Ibid., 22.

52. Ibid.

53. Ibid., 23.

54. Ibid., 24.

55. Ibid., 23–24.

56. Ibid., 23–25.

57. Ibid., 24; see also Cheney, *Annual Report of the Secretary of Defense to the President and the Congress*, 77–82.

58. Ibid., 69–73.

59. J. F. Dunnigan, *How to Make War: A Comprehensive Guide to Modern Warfare* (New York: Quill Press, 1988), 373.

60. Cheney, *Annual Report of the Secretary of Defense to the President and the Congress*, v–ix.

Chapter 6

The Vision

Following are various scenarios that illustrate the application of nonlethal technology in support of U.S. interests, or those of the United Nations via a global peacekeeping force. The intent of this section is to show that in this new strategic era, the establishment of peace must be a cooperation of policy and technology. Each of the scenarios described is currently an ongoing national threat, and each has a lethal option now being applied. The nonlethal technologies suggested are just suggestions. This book makes no attempt to promote one nonlethal technology over the other, but it does show how the nonlethal option affects the situation differently than would the current lethal option. No detailed comparison of lethal versus nonlethal or nonlethal offensive versus nonlethal defensive was conducted or intended. Empirical data is difficult, if not impossible, to come by at this juncture.[1]

The first scenario discusses the offensive nonlethal use of biocontrol narcotics agriculture (BNA) in support of the president's counternarcotics strategy versus the conventional lethal means now in place. The second scenario illustrates the use of ultraviolet light directioning and ranging (UVE-LIDAR) as a passive, nonlethal intelligence tool, again in support of the war on drugs. The third scenario proposes the offensive nonlethal use of computer viruses to degrade, disrupt, delay, or destroy a hostile nation's mainframe computer networks.

Scenario four examines the use of the shroud round or "project shrink wrap," as the technology's scientists refer to it. It is an abstract of the scenario in the prologue. Scenario five explores the use of superpolymers, which are high-tech resins and glues that can glue people and weapons in place once applied (another abstract from the prologue).

Number six introduces microbic weapons, illustrating the powerful effect of microbic technology against equipment. Number seven introduces combustion modifiers into the battlefield, while number eight shows the motor march of a mechanized force with antitraction technology. Scenario nine pits a riot situation against calmative agents and infrasound weapons. Number ten shows how super-caustics can ground an air force. Scenario eleven discusses the use of hydrogen embrittlement, and its ability to kill antenna arrays and destroy metal, plastics, and anything else subject to freezing, and the destruction of metallic harmonics.

Scenario twelve employs infrasound and high-powered microwaves. Number thirteen introduces project "Flicker" as its developers have labeled it, demonstrating what a simple candle or light bulb can do at a social gathering.

SCENARIO ONE: BIOCONTROL OF NARCOTICS

In the three "Tier I" drug countries of Peru, Bolivia, and Colombia, the fight to protect U.S. interests under the president's national drug control strategy rages on. Illegal drugs continue to flow into the United States at an alarming rate despite increased efforts by law enforcement and U.S. Department of Defense personnel. This is a perplexing national security threat, which the president answers by directing a comprehensive plan to attack to flow of drugs, at the source and in transit.[2] Brigadier General Thomas Montgomery, acting director of strategy, plans, and policy in the Pentagon, sends a message providing top-down guidance for future army involvement in the president's counternarcotics plan.[3] In this message General Montgomery provides guidance pertaining to specific counterdrug issues, including a directive to research, develop, and acquire nonlethal technologies that will support Department of Defense (DOD) and non-DOD agencies in their counterdrug roles of detection and monitoring.[4]

The drug cartels in Colombia, Peru, and Bolivia continue to wage a near-perfect strategy against the U.S. counterdrug effort. Furthermore, it is estimated that the combined cartel's incomes greatly exceeds the U.S. budget for the war, meaning that they have the means to provide for any contingency. The cartels use secure radios, high-speed aircraft, and automatic weapons and employ intelligence collection assets in their countervailing strategy. Under the circumstances, it is ironic that the United States calls this a war. U.S. forces are not permitted to engage traffickers in combat, nor are they permitted to arrest them inside or near the U.S. borders due to the posse comitatus. The hands of the military are virtually tied, while the traffickers continue normal operations.

The demand for cocaine in American society is not decreasing, and the efforts of the U.S. Drug Enforcement Agency (DEA) and DOD augmentation forces fail to decrease the flow significantly. Therefore, while the conventional war on drugs continues, something less conventional is needed to impact some aspect of the drug cycle: cultivation, processing, transporting, or sales.

The most easily affected is the cultivation, but this has little impact on the cartels' monetary gains. However, it does impact on the revenue of the pheasants cultivating the coca plant for its cash value. In part, defoliants have not been used because of the potential unrest this could cause in already unstable regions. Another problem with defoliants is that their use is considered chemical warfare, which is not welcome in most nations. Plant eradication using local police forces armed with machetes and flame has minimal results because the locations are remote, in steep terrain, and in some cases well guarded.

In this situation, what is needed is a nonlethal capability to attack the coca plant yet permit its harvest, so the peasant is not directly affected, while disrupting some aspect of the plant before, during, or after its transformation into cocaine. The nonlethal technology that has the capability to do this called biocontrol narcotics agriculture (BNA).

A very provocative thinker, Pat J. Unkefer, who is a plant pathologist with the Los Alamos National Laboratory, in New Mexico, is proposing the development of a method to biologically control the production of cocaine. To do this, Unkefer will capitalize on the host-specificity of plant pathogens to develop the specific pathogens of the plant that produces cocaine. Unkefer and her team have developed a plant pathogen that is a plant growth-altering agent using a specific vector produced by a plant pathogen to control plant metabolism and growth. This technology can be applied to inhibit (or eliminate) either the plant growth or the biosynthesis of the controlled substance in the target plant. Simply put, Unkefer can use these pathogens to stunt growth or render the plant benign by destroying the psychoactive ingredient. If the latter form of the technology is applied—eliminating biosynthesis—a coca plant could be harvested, processed into cocaine, transported, and sold, but the end result would be an inert substance incapable of producing any psychostimulation.

Another capability of this pathogen is the ability to self-inoculate, thereby eliminating the need to endanger law enforcement officials or military personnel during the process of introducing the pathogen in concentration to the plants. This danger is not from the pathogen itself—as that is harmless to humans—but in risking life by exposing officials and soldiers to cartel security forces, who would surely take action against the inoculation of their crop.

The development and use of this form of technology will have a devastating effect on the cartels' drug industry. Repercussions spanning virtually every link in the trafficking chain and management network would be felt. This technology has the potential for use in a deception campaign targeting the cartels, or simply in the disruption and delay of production. If so designed, it can also be called on to completely eradicate the plant.

The use of this form of technology supplements the U.S. counterdrug effort by adding a new dimension of weaponry not capable of being countered by the cartels. The technology protects U.S. interests by attacking the source of the drugs—the plant itself and the managers of the drug trade (the cartels). The use of biocontrol narcotics agriculture (BNA) minimizes the exposure of U.S. forces and those of its allies to the hostile potentials of the cartels. As with remote viewing, and all nonlethal technologies, the concept is to minimize, if not eliminate, collateral damage and loss of human life while having some detrimental impact on the enemy war machine. The application of BNA conceptually fits that role.

One would think that the defense or law enforcement communities would jump at the opportunity to develop and employ either this technology or a variant. Its funding costs amount to a literal speck on the existing research and development budget of the defense department; at last count, it was estimated that this project could be put to use in one to two years with a minimal budget of perhaps $2 million. How much does it cost to monitor, track, and seize the minimal amount of inbound cocain our law enforcement agencies capture each year? Now add the cost of training, equipping, and paying in-country law enforcement agents, military personnel, and foreign agents to work on this problem for us. What is the cost of prosecuting and defending all the low-level operatives and soldiers working for the opposition? As you can see, the cost of entertaining a possible alternative like biocontrol begins to look more appealing.

SCENARIO TWO: UVE-LIDAR

For a number of years, the United States has been involved in an ever-escalating drug war. The mandate given by President George Bush is to stop the cross-border flow of contraband in the form of illegal drugs. This mandate prompted an initial expenditure of specially appropriated moneys, which has since increased tenfold. The war rages on, not only at the borders of the United States, but deep within the jungles of Peru, Colombia, and Ecuador, the major growing regions of cocaine. Currently, law enforcement and special forces personnel from the United States are assisted by local militia and host nation forces as they track down and destroy drug-processing labs.

The problem they face is the fact that many, if not all, of the labs are difficult to find, requiring long search-and-destroy missions, which expose the teams to cartel security forces. It is a well-known fact that the cartels' intelligence, security and communications abilities match or exceed any in place by the law enforcement and defense forces. This fact keeps the cartels several steps ahead of their opponents at all times. Certainly, what we are discussing here is only one small aspect of the total drug picture, but this is where we must begin overt measures to thwart the global distribution of illegal drugs.

As teams of the U.S. Drug Enforcement Agency (DEA), their military, and host nation augmentation presently operate, the modus is for them to search vast areas of terrain to find suspected processing labs and destroy them. Given the often-difficult terrain, bad information and vague locations, the target objective is often missed or destroyed far too late to have made any significant difference. The cartels can stay ahead of teams fumbling about the jungle floor, outcommunicating or outmaneuvering them. Often, the target's complexion will change while the teams are en route. Teams may be ambushed on their way to the target, the target may be disassembled or the drugs moved, or the cartel security forces may simply stand and fight because they know that they outnumber and outgun the enforcement teams. Regardless, if the teams could find the objectives quickly, pinpoint their locations and strike before the cartel's could react, the tables would turn. Following is an example.

Sergeant Mason wiped the sweat from his face as he approached the C-130. The engines were already roaring, blowing the hot, humid air of the valley into his face. They belched the sickening exhaust of the turbines across the tarmac, making him scrunch his face from the smell. He stood at the foot of the back ramp, watching a small team of five scientists from Los Alamos National Labs scurry toward him. They were excited about introducing their new device into the war Mason had been fighting for almost five years now. He was less enthusiastic.

Sergeant First Class Mason was a special forces soldier of fourteen years; he'd been in this country fighting the drug war for eighteen months. Usually he was in the bush, working closely with the DEA agents and local militia. He was frustrated with what he was doing, thinking that perhaps he would never make a dent in the enemy. Everything he read indicated that cocaine was still saturating the streets of his home in Albuquerque, New Mexico. It aggravated him that he did not seem to be able to stop it, that the corruption was too deep in this country for him to change or overcome it. He had seen patrol after patrol get ambushed as they closed in on an objective, or spend five days wandering the hills and jungles of this land only to come back empty-handed, tired, and disheartened. It was a lousy business, and despite his objections to another good idea by the administration, he hoped that what these scientists had brought to him would work.

The device was strapped to a platform and lashed to the floor of the aircraft; once they were airborne, it would be put into operation. Everyone took their seats on the red webbing of the interior of the plane. The scientists were not as familiar with these surroundings as Mason was; he thought it all a pain, while they still thought it exciting. The aircraft lumbered into position on the dirt runway, and the pilot stood on the brakes while he pressed the throttles forward. The four turbo-prop engines screamed, rocking the plane and spinning dust and a hail of small rocks into the field behind them. The pilot released the brakes, and the "Hercules" aircraft lurched forward, forcing everyone and everything onboard to lean toward the rear of the plane.

Once at speed, the pilot dragged the aircraft's nose off the ground and flew the craft into the sky at a very steep angle, trying to get altitude before the 3,000–foot mountain range five miles away drew too close. In about ten minutes the aircraft was straight and level, cruising at 10,000 feet above the jungle canopy below.

Mason leaned over to the lead scientist on the team; pressing the talk switch on his intercom. "What does this thing do again?" He asked, pointing to the platform.

The scientist looked up from his notebook and smiled at Mason. "It looks for drug processing labs . . . and just about anything else I ask it to look for."

"Exactly how does a machine look for drug labs? I mean, you told me that it give us a printout or something—but how does it work?" He realized he had just asked a scientist to tell him how a very intricate piece of equipment worked, and it made him cringe. He thought if he could "narrow the parameters" a bit he might learn how to build the goddamn thing. "In a nutshell" he smiled at the scientist. "I don't plan on buying one—I just want to know the simple mechanics."

The scientist gave a mock laugh, shaking his head in acknowledgment. "Okay," he sighed, "it's pretty simple, really." He began pointing out pieces of the device, half sketching them in his notebook for clarification. "You see that cone shaped part? It shoots coherent light down range wherever I point it . . . and then it starts a sweeping motion back and forth from left to right."

Mason looked at him puzzled. "Coherent light?"

"Yeah! Concentrated light—a light beam like a really low-grade laser. Well, what that light does is strike particles in the air and cause them to fluoresce. Particles of anything—dust, exhaust, smoke from fires, fumes from fuels, pollen, you name it. If it's in the air, this thing will naturally hit it with the light. Do you know what I mean when I say it causes the particles to fluoresce?"

Mason shook his head no.

"Well, in basic terms its like this; when light hits the particles it causes them to get excited. They vibrate, giving off energy in the form of another light—and that's fluorescing. Think of it as fluorescent paint. Now, you see that telescope-like thing attached to the side of the cone? That's a telescope." He grinned, thinking it sounded funny. "What it does is look at the particles that are giving off this light energy and feed the image into that computer there on the platform. The computer is trained to recognize certain forms of light, and it matches the light to the particle and eventually to the source of the particle. Naturally, there are millions of possibilities for particles out there, so we have to teach it to only look at or recognize certain ones. Following this?"

"Yeah . . . yeah, I am." Mason nodded. "So, you can tell this thing to look for car exhaust or campfires, right?"

"Exactly! But what we are asking it to look for are fumes from the precursor chemicals used in the processing labs."

"Really?" Mason said, with a surprised look. "So what does that do for us?"

"The computer not only recognizes those types of particles, it is tracking our flight path across the ground. When it sees the plume of a processing lab—something you would not recognize with the naked eye, which you most likely could not tell from a brush fire, a house fire, or a camp fire—the LIDAR recognizes it. Best of all, it will plot the location of the plume's source down to the nearest ten feet and give it to you on the map printout. Great, huh?" The scientist stared at Mason with a wide grin on his face.

"Really great. That means we can direct a raid to the exact point on the ground within minutes of receiving the printout . . . and nothing can change. We can save a lot of time and lives with this." He paused for a moment. "Wait! They'll just move everything underground."

"Doesn't matter if they go a mile underground." The scientist countered. "They have to exhaust the fumes from the processing somewhere, and when they do, we'll find them. Even if they scrub the exhaust, they can't get everything out, and we can see down to the smallest amount of saturation or particles per any given volume of measurement. This machine's a very powerful tool. It has a lot of other military applications as well, but this is its latest adventure."

The pilot signaled over the intercom that they were nearing the target valley and that he was descending to the desired altitude of 1,000 feet. The crew chief in the rear of the plane checked all the device's tiedown straps and lowered the rear ramp of the plane exposing the jungle below. It was like looking into a huge television screen at the rear of the craft, a giant ten foot by ten foot opening spilling light into the dark fuselage. The scientists all took their positions on the device, with each checking and doublechecking their assignments. The team leader applied power to the device, which churned and blinked alive. The laser gun spun slightly and then swept itself from side to side, looking for the sides of the door—actually, calibrating its sweep for the operation.

"It's up!" bellowed the team leader. "It's ready to go!"

"Two minutes to target." The pilot squeaked over the intercom. "One minute to target . . . commencing target run."

Mason watched in amazement as the LIDAR began sweeping rapidly from side to side. He could see nothing being emitted from the gun, but the computer began spitting out small legal sized map sheets with quarter-inch red squares on them. The team leader snatched one from the printer and presented it proudly to Mason.

"Unbelievable." Mason said astounded by the product. "All of these red squares are . . ."

The team leader interrupted, "Processing labs."

"So many." Mason rubbed his forehead. "There are so many."

"Yeah! There are. But I can build up some data after awhile and tell you what quantity of base or paste is being produced at each lab—and you can pick the biggest ones to hit. How's that?"

"Unbelievable. . . . When can you start? When can we get this thing here permanently? This thing will save us a lot of time, money, and lives. We can hit these guys before they reinforce or move. Hell, we can plan an operation (op) in advance, knowing what's out there and what to expect when we get there—and it doesn't cost us a human life to find it out. We don't have to pay anybody off, the militia doesn't have to question anyone for information—we win with this thing all around. When do we get it?"

The smile left the scientist's face. "I don't know. The government is $1.9 million short in its funding for this project now . . . and most of the budget is going for other things. You know the deal—this isn't a weapon, so it isn't going to get as much attention as we'd like it to. But you keep pushing it through our channels, and we'll keep hoping we get funded. Our goal is to make this thing man-packable at some future date so you can carry it with you into the field. There are a lot of possibilities—we just need some final funding to make it happen."

Mason sat alone for the flight back to the airstrip. He needed this device; his men needed it. Why the hell didn't the government that put him here to stop the flow of drugs see that? That night he lay in bed, thinking again that he was not going to make a difference; he lacked the tools to win this war—and it would be a long time before he and his men obtained them. He would end up like everyone else, simply putting in his time and moving on.

SCENARIO THREE: COMPUTER VIRUSES

High-technology weapons continue to be available in alarming quantities on the international market place. Some Third World countries possess large, sophisticated conventional forces with tanks, artillery, air defenses, tactical air forces, antiship cruise missiles, and modern diesel submarines. The proliferation of nuclear, chemical, and biological weapons and the missile technology for long-range delivery systems will make regional conflict increasingly destructive and lethal.

At the same time that the Soviet threat is declining, the potential for major regional threats to U.S. interests is growing. Such regional threats can arise with little warning. These contingencies have always represented a variety of challenges—from Korea to the Persian Gulf, various crisis locations in Latin America, the Asian subcontinent, and elsewhere. Today such crises are made more dangerous because of the proliferation of advanced weaponry, including weapons of mass destruction, and the willingness to use them. The United States, therefore, must maintain its capability to respond to major regional contingencies. It continues to engage in arms control talks as part of a coordinated effort to enhance its security and that of its allies, but not as an end itself.

Nonproliferation and technology security vis-à-vis Third World nations in pursuit of nuclear technology, chemical and biological weapons, or missile and other strategic technologies, pose a serious threat and must be monitored carefully. The United States, it allies, and other friendly nations are closely examining this threat with a view toward strengthening their export control policies. Despite its collapse, the Soviet Union and a number of other nations continue to attempt to obtain sensitive technologies that are controlled by the United States and its allies for national security reasons.

One important national security concern is space, and the ability of much of the world to place weapons in it and wage war from it. To counter this potential, and the possibility of a ground launched nuclear exchange, President Ronald Reagan introduced the Strategic Defense Initiative (SDI) in 1982. SDI never came about, but a newer version of the concept did, called Global Protection against Limited Strikes (GPALS). GPALS is a downsized version of SDI and serves as a pre-positioned space shield designed to protect the United States and its allies from intentionally hostile or accidentally launched ballistic missiles. The system has virtually rendered nuclear weapons impotent and obsolete.

As envisioned, the defense relies on a network of spotter satellites, advanced ground-based missiles, and space-based rockets that destroy ballistic missiles through kinetic energy—the force of impact. According to William C. Loomis, director of defensive missiles systems at Lockheed Corp. (an SDI contractor), "The technology is well understood, available, and could be deployed in the mid-'90."[5] Following is a hypothetical scenario in which a nonlethal technology is used to prevent initial confrontation and escalation to nuclear war.

In 1994, the government of India entered into secret bilateral negotiations with the United States to purchase the Global Protection against Limited Strikes (GPALS) space system. While the Indian system would only serve as a small part of the overall defense net, the United States considered foreign sales to be a necessary transition enabling allied governments to provide for their own protection. The government of India assumed control of the system after the U.S. Department of Defense and NASA launched and positioned the system satellites. Two years after taking control of the system, peace alliances between the government of Pakistan and India began to deteriorate, and India threatened to use GPALS in an offensive role against Pakistan.

The U.S. National Command Authority began to weigh the options. Option one may have been to employ U.S.-operated GPALS against the Indian system. However, that option was ruled out as too dangerous, as under the circumstances, the Indian government might react to the U.S. intervention as hostile and escalate to a nuclear exchange. Another less aggressive, but equally effective, option is to employ a nonlethal, artificial germ known as a computer virus. The technology focuses on anything functioning with a computer; fire control systems, data management and transmission, fire control systems, avionics, and so forth. The virus can be engineered to fit any number of delivery mechanisms or other parameters. It can exist as a "logic bomb," a simple disruptive virus, or as a "worm," which may remain dormant and hidden until the system is used or meets the predetermined parameters. The virus can be installed at critical points in the manufacturing process of the affected weapons system—or it can be microwaved into the system from a tremendous distance.

The entire GPALS system operated under, and was controlled by, computers. All the computers are networked, and the introduction of a destructive virus would disrupt and, if desired, destroy the government of India's ability to engage the GPALS system and issue firing commands.

This virus would buy time for diplomatic mechanisms to take effect, and if the situation continued to degrade, the United States could escalate the nonlethal option to include an attack of all India's national computer networks—shutting down transborder data flow from world banks, communications, marketing, media, and so forth. By inoculating the national computer networks of India, the United States could impose sanctions by nonlethal means that would quickly cripple the country.

If this scenario sounds far-fetched, consider the following. When the possibility of a computer virus was first mentioned in 1984, nobody took it seriously. It did not take long, however, before the first wide-scale computer virus infection swept the United States, in 1986. To date, thousands of businesses and government offices have suffered from virus contamination. Unlike older viruses of the 1986–1987 vintage, the new viruses are highly destructive and can be programmed to destroy data on hard disks, corrupt files, and deceive operators of computer hardware by falsifying screen images and data translation—and they are becoming more lethal and destructive every day.[6] Whether benign or malignant, computer infections spread surreptitiously from computer to computer via telecommunications networks, hidden with other software on shared diskettes.

Bob Hein, a project engineer at the Army Center for Signal Warfare, said, "We're in the weapons business, and we're seeking a weaponized virus—malicious software that can be deployed against an adversary."[7] The army's interest in computer viruses stems from a controversial article in the trade publication *Defense Electronics* in 1989. "The application of computer virus concepts," wrote electronic-warfare expert Myron Cramer, "has made possible the advent of a new class of electronic warfare."[8]

U.S. News and World Report recently published a book called *Desert Storm: War Without Victory*, which claims that the United States employed a computer virus against Saddam Hussein's command and control computer system. A memory buffer chip was inoculated with a preprogrammed "worm" virus. This chip was then placed inside a printer and smuggled in via unwitting agents from France. Once activated, the virus disrupted the computer system by attacking all systems linked to it in the network. The virus was programmed by the National Security Agency (NSA), in Fort Meade, Maryland, and the operation was controlled from that location.

Skeptics claim that the opening of this technology would begin a new form of arms race. Richard D. Pethia, coordinator for the Computer Emergency Response Team (a high-tech special weapons and tactics team the Pentagon set up in 1989 to help with combat virus outbreaks), asked, "Do we really want to escalate a war where we are more vulnerable than anyone else?"[9] While his apprehension is understandable, it is based on the assumption that Europe, Third World nations, and others will not soon tie large amounts of their data management to computer networks. I would suggest that this is a shallow perspective, however. There are few, if any, national banks, corporations, governments, military hardware, or weapons systems that do not require some sort of computer interface. Furthermore, the global network of computers and international data flow is growing much faster than Pethia seemed to imagine.

This technology offers interesting nonlethal possibilities in support of national interests. The high-tech world of weapons, economics, and geopolitics is ever more closely tied to electronic gadgetry. The ability to delay, disrupt, destroy, or manipulate these systems via computer viral warfare is an investment in security that the United States should not overlook.

SCENARIO FOUR: SUPER-POLYMERS

The sun had not yet risen, but the smoke rising from the cookfires of Maglaj could be seen clearly in the beginning morning nautical twilight (BMNT). High on a hill overlooking the waking city, a Serb military commander pulls the binoculars from his eyes and, in a low voice, speaks to his radio operator: "Tell them to open fire."

Two kilometers away, six Serbian howitzers cough in unison sending explosive shells high into the air to fall on the innocent city. The radio cracks, "Shot over," and the commander again places the binoculars to his eyes. Searching, searching—suddenly the radio operator relays the words of his counterpart at the gun position; "Splash over." "Three seconds to impact," the commander whispers to himself, his eyes still searching.

Instantly, the building erupts in an ear-splitting thunder. Eight-year-old Kardos is thrown from his bed by the blast wave, and everything in his room falls to the floor as if thrown by some unseen being. Shards of glass penetrate his bedclothes and cut him so quickly it's painless. Outside the apartment, a young couple walking to work are so close to the shell's point of impact that they are vaporized—all that remain are small pieces of charred flesh.

The shells continue to fall, and the terror mounts. The last one was close. Where will the next one be? One after the other, the shells hit, each preceded by the sound of a ripping sheet, then a moment of silence, followed by an unearthly THAWAAK! The screams outside rise in a crescendo as plaster falls from the ceiling in Kardos' room. As he has done so many times before, Kardos makes his way, bleeding, under his bed. His world grew quiet again, and he falls back asleep, listening to the screams and breathing the thick scent of cordite.

The Serbian commander checks his map for the coordinates of his new target. "Fire mission," he squawks to his radio operator. The operator robotically repeats the command and coordinates into the radio handset. At the gun position, the fire direction center (FDC) plots the commander's coordinates and computes firing data for the guns. The FDC calls out the new data, and the gun crews spring into action, elevating the tubes and leveling the sights on a new azimuth.

The Serbian commander mounts a large rock formation. Comparing his map to the terrain below him, he begins selecting his next target, but his gaze toward the city is interrupted by an unfamiliar, distant noise. "This is not from the city," he thinks. He cocks his head to one side, trying to pinpoint the source, then glances at his operator, who shrugs his shoulders in ignorance while squinting toward the horizon. The sound grows more intense, and the commander's heart begins beating faster. "Aircraft?" the operator asks.

The commander's heart beats so fast and so loud he could no longer hear the sound. "Get hold of yourself," he thinks, "if it's aircraft, they'll never find you—besides, they are after the gun position." The commander knew all too well the capabilities of enemy close-air support. If this was indeed an attack aircraft, then this one aircraft could well take the top off this mountain with any combination of weapons. The commander had previously been on the receiving end of an air strike—and he did not want to be anywhere near its target again. "Stay low," he barked to his radio operator as they both squatted in the well-camouflaged bunker.

Twelve kilometers away, the fast-moving object pierces the quiet sky of the valley. Jerking like a small toy plane, the craft turns abruptly to the right and descends like a hawk. The commander quickly acquires the craft and cautiously brings his binoculars to his face. He cups his hands over the ends of the binoculars, being careful not to let the early morning sun reflect from the optics and compromise his position. "What the hell is that?" he asks aloud. His operator responds, "I don't know," feeling stupid after he says it, as he could not even see what his commander is looking at. "It's some kind of aircraft," the commander mutters, "one like I've never seen before . . . flying very erraticly."

The odd craft homes in on his gun position with almost uncanny precision. The officer in charge of the howitzers cries out, "It's a cruise missile—take cover!" The craft drops to a height of 50 feet and levels off on the target. Closing the gap at 400 knots, it pushes its swept-back wings forward to stabilize for slow flight. Air brakes flare out around the fuselage, and its speed abruptly decreases to 150 knots.

Five hundred meters from the gun position, two nozzles spring out of the aft compartment of the craft. One hundred meters out, a yellow cloud of vapor begins pouring from the slow-moving object. It passes over the gun position without so much as a sputter. Climbing up and away, the craft turns for another run at the position.

"Oh my God!" screams one of the Serbian sergeants, "They've gassed us." In his earlier years, he remembers being taught that nerve gas came in the form of a yellow cloud—and as this cloud slowly descends upon his position, he is convinced that agony would follow. As the vapor settles, he hopelessly holds his breath. As the substance hits him, he feels his skin tighten as if it is going to tear away from his frame. His clothing solidifies as though he had been dipped in liquid bronze—and the drone aircraft makes another pass.

In a few minutes, it is all over. The aircraft moves away as quickly as it had come—pressing its wings back against the fuselage, it gathers speed and disappears over the ridge line. "Goddamn it! What the hell is this?" The Serbian soldiers try to pry themselves from the ground. Several had unconsciously held on to equipment as they sought cover and now struggle to free their hands from the metal of their guns. "Damn!" one exclaims, making a futile effort to free his arms from the sides of his head. He had put them there to protect himself from the blast, and now they seemed to be there permanently.

Within a few hours, the soldiers will free themselves from most of their predicaments. The yellow cloud of substance with which they had been covered was not chemically engineered to adhere to skin or other organic substances. However, this was not the case with their guns and shells. Every horizontal surface had been covered with a quarter-inch film that would not come off. Despite their best effort, no solvent they possessed would soften this stuff.

Their shells were permanently stuck together, as were the breech doors of their guns. Several of the men beat feverishly on the guns' traversing handwheels, trying to free them so that the elevation and direction of the tubes could be corrected. One by one, the handwheels shattered under the blows of hammers.

Nothing worked. The carriage wheels were frozen so steadfastly that the guns could not be moved from their positions without a crane to lift them out. Every piece of equipment, no matter how large or small, was rendered ineffective. Even the men's coated uniforms had to be removed so that they could move in some fashion other than slow-motion. In frustration, they walked naked from the gun position, unable to fire and unable to repair the damage inflicted by that "Goddamn thing!" Kardos would sleep uninterrupted tonight.

SCENARIO FIVE: NONLETHAL BALLISTICS—THE SHROUD ROUND

In another part of the world, soldiers lie in wait, watching for signs of enemy movement from their antitank position. Even though they are heavily dug in, the hot desert sun bears down on them in their open trenches, scorching the backs of their necks and heating the barrels of their weapons to untouchable temperatures.

Over the desert horizon, an invading tank army motor marches toward the dug-in defenders' border at 50 miles per hour. The tanks number 150, and once in attack formation, they will stretch as far as the eye can see. The sophisticated target-acquisition systems of the tanks constantly hunt to the front of the moving army. The scanners are capable of detecting the stationary movements of even a small animal. They probe the horizon, moving, searching for a target.

As these two great armies prepare themselves for the ultimate clash, another, smaller, force moves into place, undetected by the belligerents. The troops' movements are cloaked by an electromagnetic screen as they select the best firing positions for their long-range weapons. A young lieutenant quietly directs the placement of his weapons systems on the ridge overlooking the antitank defensive positions below him. The blue helmets of his United Nations troops glisten in the sunlight. Were it not for the cloaking (something that still made him nervous because he did not fully understand it), his position and movements would surely be detected by one or both of the armies below him.

"Forty kilometers and closing rapidly, Sir," his platoon sergeant blurts out. The lieutenant's pulse quickens, much as it did back home before an opening football kickoff. "Quickly men—get these damn weapons into position,—they're coming quick." Each section sergeant moves with a purpose, barking at his subordinates until each of the ten weapons is positioned in such a way as to provide the optimum shot at the army of moving tanks.

"All weapons in position . . . communications are up and functional, Sir," the platoon sergeant tells the lieutenant, who responds, "Fine, all we have to do now is wait—I hope."

Minutes seem like hours to the lieutenant as he watches the monitor of his long-range target scanner. "I'm always amazed that their sensors can't detect these passive scanners," the lieutenant thinks as he watches the tank army close on the defensive positions below him.

His scanners were different from those of the closing army. They did not produce an electronic signal or "footprint"; rather, they operated in a way analogous to a large sponge, collecting optical data from the battlefield and digitizing it on the monitor for easy reference. His monitor displayed the speed, type vehicle, direction, and templated attack formations of the tank army. Because his scanner did not produce a footprint, he could not be seen by the approaching army's active sensors.

"Twenty kilometers—it appears they are moving into battle formations, Sir." "Thank you," replies the lieutenant. He carefully adjusts the boom-mike on his headset so that it gently touches his lips. He feels for the push-to-talk switch, and holding his hand there, he takes a deep breath and stares into the monochrome screen.

The lead tank commander's sensors begin to detect the dug-in positions forward and to his right flank. He radios to his subelements to "button-up," and each tank commander disappears into his vehicle, dropping the heavy hatch behind him. Their vision now obscured, the tank commanders must rely on computers to be their eyes.

As the tank army closes the distance between its forward elements and the defenders, the tanks' sensors are detecting, pin-pointing, locking in, and prioritizing targets. On-board computers look omnidirectional and resection the multiple azimuths—feeding digital coordinates to the tanks' main gun. The tanks' computers begin talking with one another, sharing information to ensure that no two of them would attempt to destroy the same target.

"Ten kilometers," the lieutenant tells his gun teams over the intercom. Below him, the defenders squat in their bunkers and stare through their gun sights—waiting for the tanks to get close enough for a line-of-sight shot. "The poor fools would never stand a chance," the lieutenant thought. "Nine kilometers [within range]—fire at will," the lieutenant screams into the mike.

In an instant, five howitzers send projectiles accelerating toward the tanks at twice the speed of sound. As quickly as each cannon is emptied, the electronic loaders purge and fill each breech with another round. The weapons fire another volley as the first rounds impact their targets.

The lead tank commander's vehicle is struck by the first volley. Microseconds before the round impacts his tank, it explodes into a wide opaque screen. Like an octopus enveloping its prey, it surrounds his tank with a wire-reinforced polymer film. His computer systems are blinded by the metal wire; even his infrared systems can no longer see. "What's happening?" he screams at his driver, "What the hell is happening?"

The driver feels the tank's momentum slowing as the turbine engine's rpms begin dropping. He presses hard on the accelerator, but the huge tank grinds to a halt. He tries violently to pivot steer the tank to the right or left, trying to free the vehicle from whatever is smothering it. He peers through his periscope, but he can see nothing.

In a panic, he unlocks the driver's hatch. Maybe if he can see outside he can maneuver away from whatever is gripping his vehicle. He presses upward with his arms, but the hatch is frozen solid, held fast by the polymer coating. He unfastens his lap belt, slides down in his driver's platform, and puts his feet on the hatch. He presses with all his strength, until he thinks his eyes are going to pop out of his head—but the hatch does not budge.

Tank after tank is shrouded in polymer film and dragged to a standstill. The tank commanders try in vain to communicate with each other, but as each tank is enveloped, the aluminum wire mesh in the film shorts out any incoming or outgoing transmissions. The tanks are dead. They cannot see, they cannot communicate—they cannot kill! As the air to the turbine engines is shut off by the film, the engines die from lack of oxygen. Those escaping this fate are quickly shut down by the crew because the exhaust ports of the vehicle are blocked and the cabin fills with toxic exhaust fumes.

The second echelon of the tank army does not penetrate the carcasses of its first echelon. Bewildered by the unexplained engagement of its lead echelon, the army commander directs the remainder of his force to return to their original attack position 80 kilometers to the rear. In a cloud of dust and fumes, the army turns toward their homeland. Behind them, their crippled first echelon waits for whatever fate has for them.

The tank commander keys his mike, "Whisky five niner . . . this is Romeo five niner, over . . . I say again, Whisky five niner . . . this is Romeo five niner, come in—over." There is no answer, nothing but static. The tank commander switches off his radio, and waits. The temperature in the silent tank grows rapidly, and the men of the crew look to the commander for answers, but he has none. They are prisoners in their own war machine.

The lieutenant, looking down from his vantage point, directs his radio operator to switch radio frequencies to the command net of his superior. "First echelon stopped, I estimate about thirty-five to forty; the second echelon has terminated their attack and appear to be returning to base." "Roger," bellows the voice on the other end of the radio, "enemy prisoner of war teams are en route to the tanks." "We have contacted the commander of the ground defenders below you. He has agreed to disarm, and vacate his positions." "Well done, lieutenant."

SCENARIO SIX: MICROBIC TECHNOLOGY

We have seen a variation of this technology in use throughout the world in assisting the cleanup of major oil spills. As with naturally occurring plant pathogens in the biocontrol scenario, these microbes are specifically engineered to attack certain materials with predictable results. When used in conjunction with microencapsulation technology, the result can be programmed or timed to occur when conditions are optimal or the result most catastrophic. The following hypothetical scenarios portray attacks on fuel supplies and internal components of turbines and other internal combustion engines. These microbes can be introduced into standing fuel supplies by any number of vectors: hand, projectile, aerosol, and so forth. Once introduced to the host fuel supply, they can, and will, degrade the fuel by digesting critical molecular components, causing an incomplete burn of the fuel by a vehicle's engine. Picture the result of jet fuel contamination: the result is the grounding of an entire squadron due to ground engine failure or insufficient power at run-up or takeoff. The results would be devastating to the contaminated force. These microbes are virtually undetectable unless you know what to look for and are willing to explore and test entire fuel supplies to find them. Some of the microbes are self-inoculating, passing through generation after generation while existing in the fuel supply. To detect the microbe, one would have to time the testing to match the growth and maturity cycle of the contaminant.

Another brand of microbe has been engineered to attack and consume filter components within the engines of the targeted vehicles, destroying nylon, rubber, cork, and paper gaskets to such a degree that the lifeblood of the vehicle pours out on the ground in the motor park or in a tactical field position. Microencapsulation can be programmed to release the contaminant at predetermined temperatures that are met only when the vehicle is at full operating temperature (e.g., on a motor march or battle formation). The latter makes the logistics of recovery and repair extremely costly in terms of time, material, and man-hours. Enormous numbers of vehicles and aircraft would be rendered incapable of their mission through the use of microbes to contaminate fuels—without ever firing a shot, a fire mission, or a bombing raid.

There are microbes that can attack propellants and would render an enemy's artillery rounds, or any indirect fire munition relying on charge bags or exposed propellants, insufficient, unreliable, and in most cases, unusable. These microbes consume the propellant, leaving a by-product of digestion that makes their presence virtually undetectable until the propellant is used. Rounds will fall short, rendering ballistic data in the computerized firing-direction center worthless; other weapon systems will never get a projectile to leave the tube.

Solid fuels for rockets, cruise missiles, multiple launch rocket systems, and even ICBMs are subject to contamination by these microbes, as the vectors are too numerable to adequately protect against contamination from all possible sources. Further, it may not be known whether contamination has already taken place since encapsulation may have been used. The process of tear-down and fuel inspection would wreck havoc on virtually every weapon system in the inventory that uses a combustible fuel.

Envision plaguing the Iraqi Army's fuel supply during the recent Gulf war. Imagine for a moment the destruction of their war machine immediately, and over time, if the coalition forces had taken steps to destroy their army from within using microbic technology. Skeptics will surely choose to not believe that microbes can do what is portrayed here, or the suggested use might be written off as impossible because it smacks of biological warfare. The major concern would be that microbes open the door for more severe forms biotechnology to be used in retaliation. The logical response is that if the Iraqis could have done so, they would have. However, once we had destroyed their delivery capability, they would have been unable to use it. I suppose they could simply open a valve and hope the prevailing wind drifts toward the target. It is unlikely that fear of reprisal with a duplicate technology kept that from happening. It was most likely lack of ability rather than lack of desire or resolve that kept the alleged chemical and biological weapons in their inventory from being used.

Still other microbes can turn fuel into immovable blocks of semisolid gelatinous fuel, forcing them to be left in place. Other variations can make the supply so corrosive that hoses used to pump it and composite and metal tanks used to store it break down and dissolve while holding it. The latter is not environmentally sound and would likely be a last resort if used. The beauty of all of these technologies is that the fuels can, in many cases, be reconstituted by teams possessing the technology to do so. Naturally, countermeasures will eventually come into play, as will strategic and tactical deception practices, but the end result remains significant—with minimal environmental damage, no loss of life (hypothetically), major equipment damage, and complete destruction of the fuel-consuming war machine.

SCENARIO SEVEN: COMBUSTION MODIFIERS

Combustion modifiers are chemical in nature and are designed to destroy engine components directly or choke engines by blocking their ability to utilize the oxygen necessary for combustion and power. They can be introduced into the battlefield via foams, aerosols, or liquids. They are most often presented as external contaminates that are ingested by the engine, where the damage is then realized. They are not microbes, as discussed in Scenario Six; however, they can be placed in fuel supplies where they are ingested directly into the engine and the destruction occurs. In the aerosol form, these modifiers can be inadvertently inhaled by human beings, just as the machines do. One would suspect a danger to humans encountering such a situation, but according to information available, there is more of a health hazard from inadvertently breathing the exhaust fumes of the vehicles than in breathing the aerosol modifiers.[10] A hypothetical scenario follows.

The commander of a small United Nations peacekeeping force is trying to protect a small, unarmed village from a forceful takeover by a hostile opponent. He has limited fire power and knows that he would stand little if any chance in a direct confrontation with the mechanized foe. His rules of engagement do not permit him to engage hostile forces unless he is fired on, which under the circumstances is unlikely. Understanding the U.N. commander's limitations, the opposing commander normally moves directly past or through his opponent's positions and roadblocks without ever firing a shot. In this way, he is held blameless in the eyes of the global community while still capturing and holding ground with his forces.

This time, the U.N. commander has something up his sleeve; hopefully, he can even the odds. He has been supplied with a small special munitions force (SMF), who are now under his operational control. In a late night briefing, the SMF team leader tells the commander what his force can, and will, do for him.

"The approaching force will obviously look for the most open and high-speed avenues of approach as you know. They will have to cross several ravines and bridges in this area, here, and here." He points to the locations on the map. "What I can do is stop them from crossing any of these points or areas."

"For how long?" The commander asks.

"Indefinitely! You tell me how long, and I'll prepare for it. I may need a resupply of certain substances in a day or so if it rains or the wind gets out of hand. Otherwise, we'll litter the place with his dead vehicles, and he won't know what happened."

"How can you do this with these chemicals you have?" The skeptical commander asked again.

"It's simple really, Sir. We will place persistent foams in these choke points here and here; on the bridges and between the hills on these narrow roads. In the low ground we will employ electric fog generators that will bellow a chemical mist in the hollows and open fields. Anything crossing the bridges will ingest the foam and instantly choke on it, stalling the vehicle in place. Since it's on a bridge, the bridge will soon be rendered impassible. If they try to bypass it and circle around into a lower crossing point, the mist will kill their vehicles in place."

The commander rubbed his chin for a moment, pondering their plan. "Can't they restart the vehicles on the bridge after awhile and drive away?"

"No Sir! They can't. You see these foams and mists are laced with a corrosive mixture that will actually seize the engine; that is, they are so corrosive that the engine actually melts its internal metal parts together. There is simply no lubrication, intense friction, and tremendous heat—all in a matter of seconds. It's like running them without any oil or fluids in the desert sand—only far worse. They're dead in the water, with no engines, no generators, nothing that breaths air. Except for people—the stuff won't hurt them. It gives you a little cough for a while, but it goes away after a day or so when you get into the fresh air."

"Amazing! Absolutely amazing. I've never seen anything like it." The commander was astounded.

"And how long does this last again?"

"Well, for the mist—it's obviously dependant on the weather; but I can keep up a fairly persistent fog for about 48 hours without resupply. If the wind is strong and spreads well, I can hold off on putting down a wall of mist and wait until they get close enough for us to spread it quickly across the battlefield. Either way, they won't get through."

"What recourse do we have if they penetrate the first line of defense—what else can we do to protect the city?"

"We can place foam on all the roads leading into the city—at every intersection, between every building and alley. We can form an entire perimeter with the stuff, 20 to 50 feet wide and maybe 10 feet tall. You won't believe it until you see it, and it'll stay in place for as long as we want if it's mixed to be durable."

"Can we start employment tomorrow?"

"We can begin tonight, if we agree on the targets. The drawback is that all traffic is restricted, ours as well as his. To combat that, we will sequence the obstacles in phases, leaving several open for the time being. They can be quickly closed if need be."

"Fine, Major. Please begin at once."

As the sun rose over the city the next morning, the commander stood with the major and surveyed his strange, new chemical defenses. Huge walls of foam sat on several of the stone and metal bridges leading into the city. The low ground surrounding these points was filled with a bluish haze the following morning.

"The blue haze is indicative of the type of modifier we are using. There are versions that are clear and virtually undetectable; but this is what we have for now. We may be surprised, perhaps looking at a well filled with the chemical will serve as a deterrent itself."

The defenses stood for several weeks, renewed periodically by the SMF. The opposing commander sent scouts and other reconnaissance vehicles forward to survey the positions and routes leading into the city, but they never tried to penetrate any of the strange haze or walls of white foam. Then it came; at 0200 hours, three days from the last known recon of the positions, the U.N. listening posts radioed hearing tracked and wheeled vehicles moving en mass toward their positions. The U.N. commander, feeling nervous, relayed that enemy movement was detected five kilometers southeast of the city and that he expected to be attacked before daybreak.

The first enemy tracks approached the bridges two hours after being reported. As rehearsed, the U.N. peacekeepers manning the positions drew back from their defensive sites on the far side of the bridges, leaving them exposed. A T-72 tank rumbled to a stop 50 meters from the first wall of foam. The U.N. commander had taken up a position near the bridge to observe the attack; he knew the opposing commander would not use deadly force if he thought he could bypass the positions and run his force into the city. The U.N. commander was counting on him to do just that. Two more tanks halted just short of the bridge, keeping their interval.

The lead tank buttoned up, the heavy hatch slamming down in the night air. Its engines whined as the driver ran up the rpms before releasing the brake and speeding toward the foam. The tanks struck the foam, sloshing it over the guard rails and up over the turret. Within seconds the tank began to slow. A grinding, metallic sound came from the engine, followed by a repetitive clanking until the tank coughed one final time and died on the bridge. The trailing tank bypassed the second tank, which was apparently having reservations about crossing the bridge. The trail tank sped past him and slammed into the rear of the stalled tank, trying to push it off the bridge and clear the roadway. Again, within seconds its engine ground to a halt in a painful-sounding metallic gasp. The hatches came open, and the troops poured out of the tracks. They scurried back to the remaining tank and covered its exterior as it backed away from the bridge and disappeared into the darkness of the morning.

When the sun came up, there were 32 such scenes scattered about the country side—the defenses had held.

The U.N. commander was relieved; he had held his position, protected the innocent civilians of the city, and not been forced into a tank battle or ground fight. It could have been much worse for both sides, but thanks to the modifiers, both sides had bought some time to think about the next offensive. The commander reorganized his defenses with the SMF and prepared for the next attempt to take the city. Today had been a good day.

SCENARIO EIGHT: ANTITRACTION TECHNOLOGY

There are several versions of this technology—some readily available off the shelf and others that are still in testing with promising results. One form of antitraction technology is the ability to form chemical mud. In a dry environment, using specific chemicals in aerosol form, special teams can create track and wheel vehicle obstacles by spraying the chemical polymer onto the soil and forming a sort of chemical marsh. The exact environmental impact is not known at this time. However, it appears that a half-life can be engineered into the product, rendering it inert at some determined point in time. The marsh or chemical mud would then effectively dry up. The potential use of this obstacle is obvious; as with the combustion modifier, this technology can rapidly deny terrain while at the same time claiming vehicles. Once stuck, recovery is time-consuming, and costly—and virtually impossible.

A more readily usable form of the chemical is applied to roadways, railroad tracks, the metal skids of bridges, and of course, runways and taxiways at airports. The chemicals can be applied using large sprayers, hand-painted onto the surfaces (most suitable for railroad tracks and smaller targets), or aerosol-applied using remotely piloted aircraft (drones).

Picture a supply train carrying munitions or weapons that cannot cross a mountain range or even a small incline because there is no metal-to-metal contact between the wheels and the tracks. The highly lubricative polymer applied to the tracks prevents any traction, and it cannot be washed off by rain or other elements. The tracks can be recovered for friendly use only after being bathed in a special solvent. Microscopic elements of the polymer can be rearranged, or additives can be placed in the mix that will essentially code the solvent requirement to the polymer. In other words, the solvent used to clean the track cannot be duplicated—you have to know the molecular combination or additive in order to mix the appropriate solvent, which is a more-than-adequate counter-countermeasure.

Drone aircraft can spray airfield with polymer mixtures that will deny the use of the runways to aircraft. They cannot taxi—it is like trying to drive on slick, black ice. Planes can't take off or land—they will hydroplane and loose control, tearing their landing gear off, damaging the skin of the aircraft, and running into one another. Unable to stop or control the aircraft on the ground, the facility will be useless to combat or supply aircraft. Crews servicing the aircraft will not be able to mount the larger craft to perform inspections and wing-walking crew cannot refuel; nor can cargo or equipment be loaded.

Rotary wing aircraft will experience little effect from this technology. The exception would be the crews, who will not be able to approach the aircraft for maintenance or operations. However, virtually every other track or wheeled vehicle will experience tremendous difficulty traversing any obstacle manufactured by these polymers. The same chemical will spin the tires of trucks and cars, throwing them off graded roadways, turns, and on-and off-ramps. The destruction and terrain denial would be tremendous, without the use of conventional lethal weapons.

Foot soldiers can be affected with some versions of the chemical as well. Spraying bridges, ramps, roadways, stairs and the like will make them so slippery that trying to walk across them would be literally like dancing on ice with combat boots. The time required to secure footing, walk a few steps, fall hard, and pick up yourself and your equipment to do it all over again would wreck havoc with any dismounted unit.

SCENARIO NINE: CALMATIVE AGENTS

Sensory irritants and calmative agents are variants of chemicals known as advanced riot control agents (ARCA). They are designed to knock you off our feet and incapacitate you. A type of munition that could stand in both the nonlethal ballistics and ARCA categories is the Ring Airfoil Grenade (RAG). The RAG munition is shaped like a donut, and is impregnated with CS gas in solid form. According to tests and evaluations it works far better than rubber bullets. There is another category of nonlethal weapon, termed "area-displacement and rapid confinement devices." These consist mostly of large, airbag-type devices that, once thrown in a room or alleyway, blow up and pin everyone to the wall.

These come in grenade form sometimes known as a "Smurf grenade" or in canister form. They can be triggered by remote or by simply pulling a pin, such as one would with a conventional grenade. Following is a scenario where these types of weapons might best be employed.

Somalia offered unique challenges to the multinational peacekeeping force that was charged with protecting the civilian populace from the numerous warlords fighting for control of the country. Countless situations confronted the soldiers in which the belligerent force would purposely mix itself with the innocent population to disguise their movements and frustrate the peacekeepers. The situation presented the peacekeepers with ample requirements for intervention but little conventional lethal methods of controlling crowds, enemy movements, hostage situations, and even rescue or snatch attempts.

In one such case, peacekeeping forces had surrounded a medium-sized village in which a suspected warlord and several hundred of his soldiers were known to be hiding. Within minutes of their arrival in the village, a small demonstration began to form. This situation began to escalate, slowly at first, but increasing in size and potential lethality. The peacekeepers grew concerned. Several dozen armed men were seen infiltrating the civilian demonstration. Since these men wore no uniforms, detecting them in the crowd was virtually impossible. You would know them only if you were able to see their weapons—and then it was most likely too late. Some of the men gathered hastily in small back rooms, dragging children and women with them to use as shields as they retreated into these confined spaces. In this situation, the peacekeepers were helpless, and the warlords knew it. Any introduction of force would be met with force in kind, and a bloodbath involving innocent civilians would surely follow.

Picture the same scenario where the peacekeepers put on their protective masks within minutes of the determination that the gathering was escalating out of control. A radio call from the peacekeeper's commander brings in a remotely piloted aircraft that sprays the crowd with a slight calmative agent. As the mist settles into the crowd mixed with the belligerents, everyone begins to drop to their knees, grabbing whatever they can find. The effect is not instant; people do not fall mid stride but slowly lose any desire to move, talk, or remain on their feet. They lower themselves to the ground, and within seconds they are all fast asleep. They cannot move; they lie there peacefully on the ground, breathing slowly but normally. The peacekeepers move throughout the crowd 60 seconds later, cuffing and disarming the belligerents where they lie.

Seven of the belligerents have taken hostages as expected and are cornered in the back room of a small restaurant. A special clearing team of peacekeepers arrives on the scene quickly. The belligerents have protective masks and are not affected by the calmative gas. They have stacked the sleeping bodies of their hostages about the room; some of them have covered themselves with the bodies, hoping not to be found by the peacekeepers. A sergeant on the peacekeeping team directs his team members to fire several rifle grenades into the restaurant. The rounds are directed through the doors and windows of the rooms where the belligerents are. In seconds the rounds burst forcefully into large airbags that snatch everything and everyone in the room and slam them against the walls. The contents of the room, including the belligerents, are stunned and helpless; they cannot move from the position in which they were pinned to the wall. They cannot use their weapons; they can offer no resistance. One by one, the special team deflates the bags, rushing in to subdue the belligerents as they fall off the walls, stunned and fatigued.

Outside on the village streets, other peacekeepers are still loading trucks with unconscious belligerents and hauling them away to an internment and processing facility. The confiscated weapons are safely trucked to another destination before anyone begins to move. Within thirty minutes, some of the people begin to stir on the ground; within 45 minutes, 50 percent of the affected population are on their feet. The only side effect is a slight dizziness that will last for another six hours or so; there is nothing permanent. Within an hour, the villagers are back to normal, most likely laughing about the sudden absence of the belligerents.

SCENARIO TEN: SUPER-CAUSTICS

Super-caustics can be millions of times more damaging than hydrochloric acid. They can arrive at the target as liquids, powders, jellies, or a microencapsulated aerosol. On contact, the substance will burn, etch, scar, and eat its way into the target; destroying optics, metals, composites, canvas, and other materials. As with any of the chemicals mentioned here, it, too, can be engineered not to destroy organic material such as human skin. The bottom line is that everything in existence has a molecular code, which tells the nonlethal weapons designer what will and will not destroy the target material. Armed with this knowledge, these technologies can be made environmentally safe, safe for humanity but highly destructive to designated materials.

Super-caustics would be employed against aircraft, missiles, vehicles, munitions, buildings, and thin-skinned materials such as floating bridges, boats, radar dishes, and microwave towers. The damage is irreparable, requiring the replacement of the entire system under special handling conditions. The half-life of the caustic can be engineered to meet specific needs, giving it a destructive effect lasting from hours to weeks. It can be time released to permanently contaminate an area, destroying material over an extended period of time and destroying new systems as they replace old ones. Certain additives make it persistent, meaning you are unable to actually clean, replace, or repair the affected equipment. It cannot be touched without contaminating the personnel and equipment attempting the recovery and repair, with negative consequences.

In a situation where two potentially warring factions are preparing to face one another, when diplomacy has broken down and the peacekeeping effort is no longer effective, this technology might be used to quell hostilities and buy time for another solution to take effect. Let us assume that both sides in this growing disagreement have fairly large air forces. As total war approaches on the spectrum of conflict, it is probable that an air war or limited exchange will begin. After the first exchange, it is unlikely that the momentum will slow and likely that the conflict will escalate quickly to total war.

Before this happens, a peacekeeping effort intended to destroy the war machine, slowing the momentum and potentially ending rapid escalation, would be appropriate. In this scenario, super-caustics could have a role. As the opposing air forces prepare for combat, peacekeepers may send in drone aircraft to systematically blind ground radar stations with caustics. Other targets would include aircraft themselves. The thin skin of jet fighters and bombers would be eaten away, rendering them useless. Drums of fuel, fuel bladders, antennas, and ground service vehicles would be destroyed. Asphalt and concrete surfaces would be damaged, etching the surface and creating an inch of dust and debris.

Secondary targets might include concentrations of armored vehicles and towed and self-propelled artillery, each suffering irreparable damage. Special ground teams might attack artillery pieces, pouring caustics down gun tubes or painting a thick concentration of the caustic on armor plate, tires, and any other unexposed machined surface. The results are devastating, and the time and cost of recovery from such an attack are unimaginable. The armies and air forces of the belligerents are now a nonentity. Any weapons unaffected in the initial attack would be damage using one of the other technologies previously discussed. In due time, nonlethal technology would destroy the ability to wage war by conventional means.

SCENARIO ELEVEN: HYDROGEN EMBRITTLEMENT

Hydrogen embrittlement is a highly experimental yet effective technology. How I learned about its nature and intended uses was very secretive.[11] What it is is encapsulated liquid hydrogen. What it does is embrittle metals and other materials, causing them to shatter. The affect on humans is yet undetermined, and therefore it should be used with caution (or not at all from my standpoint). However, I discuss it here because I assume that it can, and will, be engineered to meet the criteria this book has established for nonlethal weaponry.

The substance is introduced to the battlefield as an aerosol, sprayed onto equipment most likely via a remotely piloted aircraft or any conventional high performance aircraft. The concept is the carpet an area with the spray, which on contact with the exposed surface of the target freezes, and thereby alters the material. If the material is touched, bumped, or struck, it shatters as a brittle glass ornament would. The effect to gun tubes is most understandable: if the upper surface of the tube is exposed and the underside is not; then the harmonic properties of the metal are altered. Every gun tube expands and collapses at an undetectable level when fired. That is, as the projectile passes through the tube and there is some expansion and contraction as the mass passes through; analogous to a snake swallowing a rat. If the harmonic balance of the tube is altered, the metal will not function properly, expanding and contracting unevenly—the result, a shattered gun tube. The potential effect on tanks, artillery, mortars, and the like would be catastrophic.

The technology also has application to tank tracks, thinly armored vehicles and of course aircraft, thin metal fuel storage facilities, and so forth. The difficulty in my opinion is that limited duration of the substance. I cannot imagine that what is freezing upon application would remain so for any length of time. However, given what I have seen in this field, that is certainly not impossible. The scientists involved in the development of these technologies know no limits, except funding and support. If they worked at it, they could eventually make it do whatever they needed it to do.

SCENARIO TWELVE: INFRA-SOUND AND MICROWAVE TECHNOLOGY

Infrasound technology has been around for a number of decades. It was tested for a military role by the Germans in World War II. Hitler had his scientists working on a large "sound cannon," which would concentrate sound waves and direct them at a target. The idea was to fire the weapon at thin-skinned bombers, breaking them apart in the sky. The footprint of the weapon was much larger than flack producing munitions or simple high-caliber machine guns, therefore, increasing the change of a high altitude kill.

High-powered microwave weapons are also not so new. They concept has been in existence almost since radar was developed. It was known that the concentrated radar beam could kill, and it was explored as a potential weapon early on. The new applications for the technology are well beyond the early designers' imaginations, possessing highly lethal as well as nonlethal capabilities. Again, I will discuss only the true nonlethal capabilities.

Infrasound weapons come in the form of frequency generators that are directional or omnidirectional. The frequency can be altered to cause nausea, vomiting, and vertigo in human beings. Bowel spasms are frequent and uncontrollable. Targeted humans cannot function in any capacity; they drop from their duty positions and abandon all mission capability. They remain in this condition until the sound generators are turned off. The law enforcement and military applications are obvious. In hostage situations or standoffs, police forces can saturate the target with devastating results, while officers with special hearing protection close on and penetrate the objective. They can take an objective, theoretically, with no casualties and no permanent damage to those targeted.

High-powered microwaves can do similar damage to humans. They can scramble brain waves, blur vision, and cause indecision, nausea, vomiting, muscle spasms and uncontrolled body movement. Under the effect of this weapon, humans are reduced to squirming masses that can be disarmed, taken into custody, and incarcerated, with minimal long-term affect. The affects are a bit more long lasting than those of infrasound, but they are not permanent and require no medical attention unless the weapon is used in its more severe range. In this case it can induce such severe symptoms that death rapidly occurs.

SCENARIO THIRTEEN: BINARY LIQUID-FILLED MUNITIONS

These weapons have a broad range of applications. Essentially, we are talking about "water balloon" and "squirt gun" technology with a military application. We can actually destroy or limit enemy weapons on the ground or in space with this technology. Not exotic as some of the other weapons portrayed in this chapter, these are effective nonetheless.

These are paint weapons, not ordinary paint; a paint mixed with additives making it super sticky, super-resilient and virtually impossible to remove. It can be painted, sprayed or splattered onto its target.

Using current laser-guided munitions, an F-16 lobs the munition into the air several miles from the target. The munition locks onto a laser beam being applied by a Special Forces ground team to an enemy tank. This process, called "painting the target," gives the descending munition its bull's-eye. The bomb rides the beam, making its own altitude and attitude corrections until it makes contact with the target. It is like flying a glider; it has no power of its own. As it descends, it can only make minor corrections to stay on the laser beam and strike the target. However, when this bomb hits its intended target this time, it does not explode—it splats! The munition is filled with a highly visible paint like liquid, that covers the tank instantly. It blinds the optics, shorts out the communications, and can even stall the engine. Additives can make the fluid do any and all the things done with the previously mentioned technologies, from polymers to caustics and lubricants. The tank is dead, but its crew is not.

The sheer beauty of this technology is that it obviously marks the tank as a hit. This is one of arguments the nonlethal skeptics use: "We won't know what has been affected on the nonlethal battlefield." In this case you will, and it has potential to be adapted into the other technologies as well, as by color-coding damaged targets.

In law enforcement situations where suspects are fleeing on foot or in vehicles, they could be shot and marked for easy tracking. The substance will not wash or wear off; it must be removed with a special solvent. It can have additives that make it glow in low-light environments or make it detectable from aircraft or with special optics used by law enforcement officials. Microscopic wafers or chips can transmit signals that can be tracked to the nearest ten feet via existing GPALS technology.[12] It can have a transdermal quality that slowly poisons the suspect, causing nausea, lethargy, or coma, without death.

Another application is in antisatellite use. The concept is to deploy high-flying drone aircraft that can track down enemy satellites and "spray-paint" their camera lenses, antenna arrays, solar panels, and so forth. Why is sending a lethal killer into space to laser, or ram, or fire some other munition at the target any better? Obviously, to the free thinker, it is not. These "paint ball" munitions offer wonderful possibilities.

SCENARIO FOURTEEN: PROJECT "FLICKER"

Project "Flicker" is an umbrella title for technologies that are designed to bring on motion sickness in humans. The degree or severity of the sickness can be controlled, and the effect is temporary once the source is turned off.

It has been determined that light engineered and tuned to a specific frequency and intensity will bring about nausea, vomiting, bowel spasms, and vertigo. About 50 years ago, when the technology was first being explored, the project developed a candle that would do these things; hence the name "Flicker." Currently, the technology is applied to high-intensity lights such as spotlights, which can be mounted on any number of vehicles, aircraft, or tanks. They can be directed at the enemy and produce the desired results within seconds, much as infrasound and microwaves do.

In a law enforcement hostage or siege condition, such as Waco, Ruby Ridge, or the John Singer operation in Utah, this technology could have been employed to achieve operational results with minimal or no loss of human life. Unless suspects cover their eyes for the duration of the standoff, they will suffer the effects of the weapon. According to technicians, it does not have to be a direct beam into the eyes; it can render the desired effect even if it is bounced off interior walls, through drapes or window shades, and so forth. Therefore, unless you remain in total darkness, you will be affected. A vomiting, retching, and writhing suspect who is unable to control his bowels or know whether he is up or down presents a much less formidable opponent to law enforcement officials. The use of deadly force in these type situations would almost cease to exist, which would be a welcome condition to most of the world.

In the military arena, the same is true. Aircrews, tankers, gunners, dismounted soldiers, commanders, logisticians—virtually anyone who uses their eyes to get the job done would be adversely affected. Whether peering through optics or with the naked eye, the effect is the same—predictable and nonpersistent.

CONCLUSIONS

This chapter has not been a technical description of widgets and contraptions. Instead, it is the first attempt to explain clearly how nonlethal technologies are used in lieu of lethal or conventional options. Most of the technologies described in this chapter do exist and have been tested and proven effective; they lie in the future only because they are competing with the conventional arsenal and its supporters.

However, nonlethal technologies are beginning to emerge from the laboratories; global opinion demands it, and they will have a revolutionary impact. The strategy and policy issues surrounding these technologies may take years to develop, but one thing is for certain, if they are not employed, we will never solve the problems inherent to their use. We did not comfortably accept the use of any conventional weapons system; they all come with their idiosyncracies, shortcomings, and supply, repair and performance issues. That seldom stops the employment of conventional weapons; it seldom prohibits the purchase of weapons and technologies to be introduced into the lethal arsenal. Why should we accept the stagnation of nonlethal weapons and technologies when we are so readily willing to accept risk in the adoption of new ones with, ever-increasing lethality?

The problems are not so much scientific as bureaucratic. If it wished, the Pentagon could find the money to accelerate the development of these technologies. There is little doubt that these weapons will find a place in the inventories of the technologically superior armies of the third millennium. Let us hope that they are adapted as true, nonlethal forms of the science. Further, let us pray that the United States and its U.N.

allies find cause to be first in the acceptance and employment of these weapons and technologies in a global peacekeeping role.

NOTES

1. For more information on nonlethal research findings, see Herbert H. Dobbs et al., "Assessment of Mission Kill Concept, Requirements, and Technologies," SPC Log No. 90–1987 (Arlington, VA: Defense Advanced Research Projects Agency, September 1990).

2. General Colin L. Powell, Chairman of the Joint Chiefs of Staff, *The National Military Strategy: 1992* (Washington, DC: U.S. Government Printing Office, 1992), 15.

3. Message number 0206, dated June 1991 (UNCLASSIFIED).

4. Ibid., p. 3.

5. R. Mitchell, "Instead of Star Wars, Little Star Wars?" *Defense Technology* (July 1991): 80.

6. J. Hruska, *Computer Viruses and Anti-Virus Warfare* (New York: Ellis Horwood, 1991), 3–4.

7. S. Schulman, "Artificial Germ Warfare," *Technology Review* 18 (October 1991): 18.

8. M. Craymer, "Viral War," *Defense Electronics* (Fall 1989): 78–82.

9. M. Lewyn, "Killer Computer Viruses: An Idea Whose Time Shouldn't Come," *Business Week*, 23 July 1990, 30.

10. This is according to my notes complied during conversations with several combustion scientists at Aberdeen, Maryland, in 1993. I did not ascertain whether empirical data had been gathered on a study of this nature or if any long-term effect was known.

11. I have read nothing about this technology in open-source literature. It was explained to me over a secure phone by a scientist at Sandia National Labs, Albuquerque, New Mexico, who was not affiliated with the project but stated that he was familiar with the technology. What is portrayed here is unsubstantiated; naturally, this man wished to remain anonymous.

12. GPALS technology is mentioned in Chapter 5. The actual method of tracking is similar to the commercial auto security and tracking technology called "LOJACK," with which police receivers can access the signal being transmitted from the wafer or chip that was splattered on a suspect or car, and follow the signal to the suspect's location.

Chapter 7

Epilogue

This book demonstrated that the weapons and forces created by nations to project global power are indisputably connected to that nation's strategic setting, security strategy, and military strategy. The volume is based on the assumption that force continues to be used in international politics despite its increased lethality, detrimental environmental impact, and potential for global destruction.

The final objective of this book is simply to recommend an alternative to total lethal war—as a maximum—or to suggest a supplement to the continued development of weapons of mass destruction—minimally. The ultimate goal should be the development of a global peacekeeping force armed with nonlethal weapons.

Following are several views about war in the current period. First, some analysts argue that weapons of increasing lethality have abolished war and that true conventional, protracted conflict is impossible. Nothing could be further from the truth, however. This line of thinking maintains that because these weapons exist, all war would be total war; moreover, because no one wants total war, it will not take place. The events of operations Desert Storm (the liberation of Kuwait), Just Cause (the Invasion of Panama), and Urgent Fury (the Invasion of Grenada) clearly show that the existence of weapons of mass destruction has not made war impossible. In fact, although the deterrence strategy of the United States has avoided major world war, the amount of violence in the world has not been appreciably reduced. More powerful and lethal weapons have not abolished war—it remains an instrument of policy.

A second view suggests that although increased lethality in weapons has not made war impossible, it must be abolished if humankind is to survive. Those who hold this theory argue that unless war is abolished, we will eventually witness a thermonuclear war that will destroy most, if not all, of civilization. This approach does not suggest how one can abolish war. Nor does it demonstrate that nuclear war is inevitable unless all violence is eliminated. However, proliferation in the Third World of chemical, biological, and nuclear weapons barbs the reality of this view.

A third, somewhat less extreme, approach contends that although conflict itself may continue, total conventional or nuclear war has become unthinkable. Total conventional or nuclear war must be prevented if humankind is to survive. Some who hold this opinion would argue that conventional war, its weaponry, and nuclear war should not be studied, for to study it is to allow that such a war is possible and acceptable, and therefore likely.

A contrary view, which is maintained throughout this book, is that we are most likely to be able to prevent total conventional or nuclear war if we carefully study the subject and its historical linkage to national security and military strategy. Furthermore, if we identify an alternative means to a peaceful end—specifically, a nonlethal supplement to the base force—perhaps, from meager beginnings, a race for nonlethal means of peacemaking and peacekeeping will commence. In any case, the threat of total war and the use of nuclear weapons will continue to have a profound effect on international politics and, indeed, human history.

A fourth, more extreme, approach advocates preventive nuclear war to produce peace, suggesting that nuclear war is inevitable, and the sooner it occurs, the better. Given the recent changes in the global political situation, the emergence of a new strategic era, and the threshold of the third millennium, this theory may seem removed from reality. However, as the threat remains undefined, the potential for protracted insurgencies and the possible use of nuclear weapons remains a viable issue. Understand that implicit in this theory is the belief that human violence cannot be ended, but that preventive war, not international disarmament or alternative (*nonlethal*) weaponry, is a means to this end.

The fifth approach combines a focus on the reality and ever-present threat of nuclear weapons and the fear that those weapons will eventually be used by rogue nation states or in a terrorist role by radical religious or political factions. Further, internal terrorism within the United States has become an emerging concern with the bombing of the World Trade Center and the federal building in Oklahoma City; and events in Somalia, Haiti, and Bosnia make clear the ever-increasing role of peacekeeping under the United Nation's umbrella. This is the new global reality—a reality with lethal consequences and, currently, only lethal options with which to resolve it.

There is a spectrum of conflict that ranges from political and diplomatic solutions to total war using lethal methods to resolve conflict (see Figure 7.1). Between these two extremes are a series of levels of power that must comprehend the mix of the elements of power and the straight-line cause-and-effect of humanism. The current base force structure is insufficient to meet the growing need for a balance of force with a balanced range of options for the CINCs. Essentially, we continue to supply the CINCs with lethal options in situations where other levels of power are more acceptable. These other levels of power require enforcement tools if they are to be a realistic means of conflict resolution.

It is proposed that nonlethal weapons augment the lethal arsenal in support of these intermediate levels of power projection. Initially, nonlethal weapons will provide the CINCs with a full range of weaponry that can prevent the escalation of conditions requiring more lethal force. Ultimately, these weapons will provide global leadership with sufficient nonlethal weaponry of such magnitude and efficiency that total war will become as unlikely as total peace is today (see Figure 7.2).

Policy implications are a major issue in considering the adaptation of nonlethal weapons; and they are not discussed in this work for good reason. The purpose of this book is to provide the vision necessary to understand the potential of these weapons and technologies and how they can relate to current world scenarios. It has not been an analysis of political, economic, and strategic considerations. Policy makes its own way once the decision to employ the technology has been made. Throughout history, weapons and forces have been designed, built, and put into place without an overbearing consideration of all the policy implications. If we are to progress as a species, global leaders cannot afford to continue to ignore the nonlethal possibilities, especially if this failure to acknowledge their potential is the result of policy implications. The world is moving too fast; too many opportunities for nonlethal intervention have already passed, and more confront us every day—and still we have no arsenal. The options we provide our military leadership remain highly lethal for the most part. We send troops into peacekeeping missions armed only with lethal options; all else is negotiated outside their scope of influence. If diplomacy fails, it fails them; they have no intermediate option. They stand watch, hoping that peace negotiations and treaties hold. If they do not, or if an insignificant situation develops, their only options are to look the other way, report the matter to a higher headquarters, or intervene with lethal force. This is an insufficient balance of options for peacekeepers, who must be able to quell disturbances by nonlethal means. They must be able to disarm or control the self-destructive efforts of a belligerent by eliminating the latter's ability to self-destruct. The destruction of the enemy's ability to destroy will control the frequency of attempts and, ultimately, the lethality of retali-

Figure 7.1
The Role of Nonlethal Weaponry in the Spectrum of Warfare

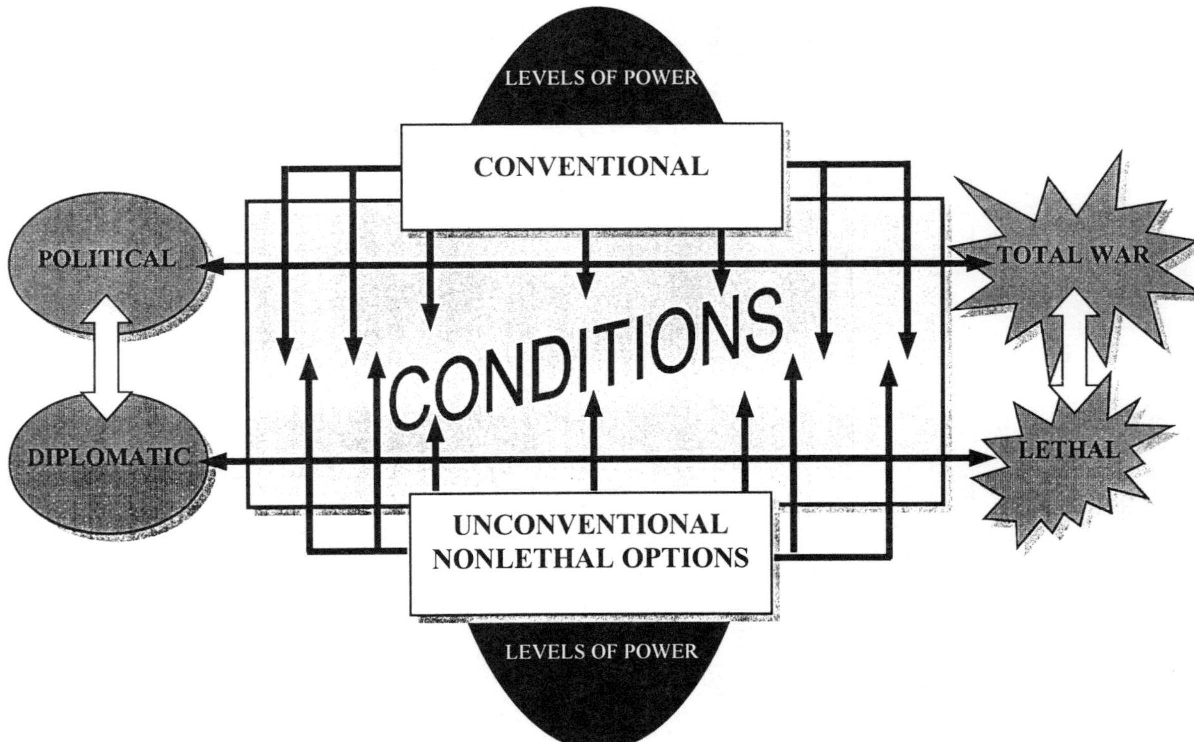

Figure 7.2
The Argument of the Global Arms Race: Why Retool for the New Millennium?

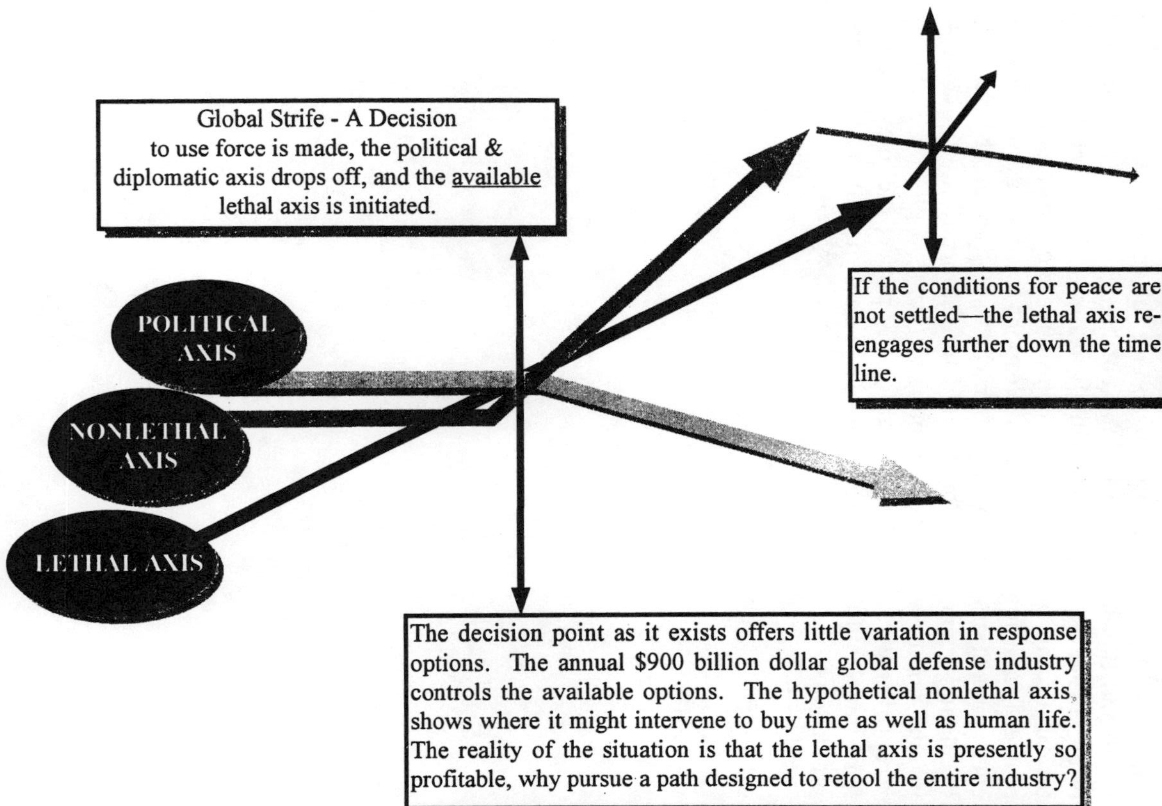

ation. In peacekeeping, the goal is to limit the effectiveness of the war machine, and in total war, to destroy it.

We will never make progress unless we take these first steps. Nonlethal weapons and technology usher in a new era of understanding. It is our key to an evolutionary step to end the prospect of mutual annihilation and the destruction of innocent life. I am not proposing the end of organized conflict, which will always exist. However, we face a historical turning point, on which will define the future of conflict as one of ever-increasing lethal options. Weapons and forces are building on an ever-increasing scale of destructive capability, absent the strategic, political, and economic rationale used to historically justify them. At this turning point, we can accept the latter, or, as a global society, we can accept that we can and must sufficiently protect the lives of our members, thus justifying the search for less lethal alternatives to achieve that protection.

Selected Bibliography

BOOKS

Alexander, J. B. 1990. *The Warrior's Edge*. New York: William Morrow and Company.

Anderson, J. K. 1970. *Military Theory and Practice in the Age of Xenophon*. Berkeley and Los Angeles: University of California Press.

Andriole, S. J. et al., eds. 1986. *High Technology Initiatives in C^3I Communications, Artificial Intelligence and Strategic Defense*. Washington, DC: AFCEA International Press.

Arbitov, A. G. (1977). "Rethinking Nuclear Deterrence: In Search of a New Basis for European Security." In F. Cerutti and R. Ragioniere, eds., *Rethinking European Security*. New York: Crane Russak, 39–41.

Bayles, J. et al., eds. 1977. *Contemporary Strategy Theories and Policies*. New York: Holmes and Meier Publishing.

Belote, J. H., and W. M. Belote. 1975. *Titans of the Seas: The Development and Operations of Japanese and American Carrier Task Forces during World War II*. New York: Harper and Row.

Blechman, B. M., and E. N. Luttwak. 1987. *Global Security: A Review of Strategic and Economic Issues*. London: Westview Press.

Boll, M. D. 1988. *National Security Planning: Roosevelt through Reagan*. Bowling Green: University Press of Kentucky.

Brown, H. 1987. *The Strategic Defense Initiative: Shield or Snare?* London: Westview Press.

Bryan, I. J. 1924. *Japan from Within*. London: T. Fisher Unwin.

Bryson, L. et al., eds. 1944. *Approaches to World Peace: Fourth Symposium*. Published by the Conference on Science, Philosophy and Religion in Their Relation to the Democratic Way of Life. New York: Harper and Brothers Publishing.

Brzezinski, Z. 1990. *Promise or Peril: The Strategic Defense Initiative*. Washington, DC: Ethics and Public Policy Center.

Carpenter, D., and N. Polmar, eds. 1987. *Submarines of the Japanese Navy*. Annapolis, MD: U.S. Naval Institute Press.

Cerutti, F., and R. Ragionieri, eds. 1992. *Rethinking European Security*. Published under the Auspices of Forum on the Problems of Peace and War. Washington, DC: Crane Russak.

Chaliand, G., and J. Rageau, eds. 1987. *A Strategic Atlas: Comparative Geopolitics of the World's Powers*. T. Berrett, trans. New York: Harper and Row.

Chayes, A. H., and P. Doty, eds. 1989. *Defending Deterrence*. New York: Pergamon-Brassey's.

Craig, G. A., and G., Felix. 1986. "Reflections on Strategy in the Present and Future." In Peter Paret, ed., *Makers of Modern Strategy from Machiavelli to the Nuclear Age*. Princeton, NJ: Princeton University Press, 863–872.

Davies, D. W., and W. L. Price. 1987. *Security for Computer Networks: An Introduction to Data Security in Teleprocessing and Electronic Funds Transfer*. Singapore: John Wiley and Sons.

Dietrich, W. S. 1991. *In the Shadow of the Rising Sun—The Political Roots of American Economic Decline*. New York: Pennsylvania State University Press.

Dunnigan, J. F. 1988. *How to Make War: A Comprehensive Guide to Modern Warfare*. New York: Quill.

Dupuy, T. N. 1973a. *Asiatic Land Battles: The Expansion of Japan in Asia*. New York: Franklin Watts.

———. 1973b. *The Naval War in the Pacific: The Rising Sun of Nippon*. New York: Franklin Watts.

———. 1985. *Numbers, Predictions and War: Using History to Evaluate Combat Factors and Predict the Outcome of Armed Conflict*. Fairfax, VA: Hero Books.

Ebon, M. 1983. *Psychic Warfare, Threat or Illusion?* New York: McGraw Hill.

Farrington, H. 1990. *Strategic Geography: NATO, the Warsaw Pact, and the Superpowers*. London: Routledge Press.

Flynn, J. C. 1991. *Cocaine: An In-Depth Look at the Facts, Science, History and Future of the World's Most Addictive Drug*. New York: Carol Publishing Group.

Foerester, S., and E. N. Wright, eds. 1990. *American Defense Policy*. 6th ed. Baltimore, MD: Johns Hopkins University Press.

Friedman, G., and M. Lebard. 1991. *The Coming War with Japan*. New York: St. Martin's Press.

Fukuyama, F. 1992. *The End of History and the Last Man*. New York: Free Press.

Gamarra, E. A. and J. M. Malloy. 1990. "Bolivia: Revolution and Reaction." In H. J. Wiarda and H. F. Kline, eds., *Latin American Politics and Development*. London: Westview Press, 359–377.

Halperin, M. H. 1971. *Defense Strategies for the Seventies*. Boston: Little, Brown and Company.

Horikoshi, J. 1970/1982. *Eagles of Mitsubishi: The Story of the Zero Fighter*. Tokyo: Kappa Books. Reprint. S. Shojiro and H. N. Wantiez, trans. Seattle: University of Washington Press.

Hruska, J. 1991. *Computer Viruses and Anti-Virus Warfare*. New York: Ellis Horwood.

Hunter, J. D. 1990. *Culture Wars: The Struggle to Define America*. New York: Basic Books.

Huszar, G. B. de. 1959. *National Strategy in an Age of Revolutions*. New York: Praeger.

Huth, P. K. 1990. *Extended Deterrence and the Prevention of War*. New Haven, CT: Yale University Press.

Iriye, A. 1981. *Power and Culture: The Japanese-American War, 1941–1945*. Cambridge: Harvard University Press.

Isby, D. C., and C. Kamps, Jr., eds. 1985. *Armies of NATO's Central Front*. London: Jane's.

Johanson, M. 1990. "Beyond Deterrence through Common Security." In F. Cerutti and R. Ragionieri, eds., *Rethinking European Security*. New York: Crane Russak, 39–42.

Kissinger, H. A. 1971. *Problems of National Strategy*. New York: Praeger.

Kitner, W. R., and R. B. Foster, eds. 1973. *National Security Strategy in a Decade of Change: An Emerging U.S. Policy*. London: Lexington Books, 431–456.

Kitson, F. 1974. *Low Intensity Operations: Subversion, Insurgency, Peace-Keeping*. Rev. ed. London: ArcherBook.

———. 1987. *Warfare as a Whole*. London: Faber and Faber, 32–36.

Kline, H. F. 1990. "Colombia: The Struggle between Traditional Stability and New Visions." In H. J. Wiarda and H. F. Kline, eds., *Latin American Politics and Development*. 3rd ed. London: Westview Press.

Kronenberg, P. S. 1982. *Planning U.S. Security: Defense Policy in the Eighties*. A Study Developed at the National Defense University. New York: Pergamon Press.

Latourette, K. S. 1957. *The History of Japan*. New York: Macmillan.

Levy, W. J. 1982. *Oil Strategy and Politics, 1941–1981*. Boulder, CO: Westview Press.

Livingston, J., J. Moore, and F. Oldfather, eds. 1973. *Imperial Japan 1800–1945*. New York: Pantheon Books.

Lorenz, K. 1966. *On Aggression*. M. K. Wilson, trans. New York: Harcourt, Brace and World.

Lusar, R. 1960. *German Secret Weapons of the Second World War*. New York: Philosophical Library.

Margiotta, F. D., and R. Sanders, eds. 1985. *Technology, Strategy and National Security*. Washington, DC: U.S. National Defense University Press.

McCloughry, E. J. 1957. *Global Strategy*. New York: Praeger.

Munro, H. 1991. *The Quick and the Dead: Electronic Combat and Modern Warfare*. New York: St. Martin's Press.

Murray, D. 1920. *Japan*. London: T. Fisher Unwin.

Murray, D. J., and P. R. Viotti, eds. 1982/1989. *The Defense Policies of Nations: A Comparative Study*. Baltimore, MD: Johns Hopkins University Press.

O'Connell, R. L. 1989. *Of Arms and Men: A History of War, Weapons, and Aggression*. New York: Oxford University Press.

O'Heffernan, P. 1992. *Mass Media and American Foreign Policy: Insider Perspectives on Global Journalism and the Foreign Policy Press*. Norwood, NJ: Ablex Publishing.

Osgood, C. E. 1974. *An Alternative to War or Surrender*. Chicago: University of Illinois Press.

Palmer, D. S. 1990. "Peru: Democratic Interlude, Authoritarian Heritage, Uncertain future." In H. J. Wiarda and H. F. Kline, eds., *Latin American Politics and Development*. London: Westview Press, 258–282.

Pfaltzraff, R. L., Jr. 1986, 1988. *National Security: Ethics, Strategy, and Politics: A Laymen's Primer*. Foreign Policy Report International Defense Publishers. Reprint. New York: Pergamon and Brassey's.

Rhine, J. B. 1953. *New World of the Mind*. R. P. Heller and M. Schindler, trans. New York: William Sloan Publishing.

Sabrowsky, A. N. 1985. *Polarity and War: The Changing Structure of International Conflict*. London: Westview Press.

Sarkesian, S. C. 1966. *The New Battlefield: The United States and Unconventional Battlefields*. Westport, CT: Greenwood Press.

Skulski, J. 1989. *The Battleship Yamato*. Annapolis, MD: U.S. Naval Institute Press.

Snow, D. M. 1987. *National Security: Enduring Problems of U.S. Defense Policy*. New York: St. Martin's Press.

Stein, J. B. 1984. *From H-Bomb to Star Wars: Politics of Strategic Decision Making*. London: Lexington Books.

Trachtenberg, M. 1991. *History and Strategy*. Princeton, NJ: Princeton University Press.

Untersher, L. 1990. "The Case for Alternative Defense: A Detail Proposal and Its Implications." In F. Cerutti and R. Ragionieri, eds., *Rethinking European Security*. New York: Crane Russak, 105–118.

Van Creveld, M. 1991. *The Transformation of War*. New York: Free Press.

Vetter, H. J., and P. R. Perlstein, eds. 1991. *Perspectives on Terrorism*. Pacific Grove, CA: Brooks/Cole Publishing Company.

Walzer, M. 1977. *Just and Unjust Wars: A Moral Argument with Historical Illustrations*. New York: Basic Books.

Weigley, R. F., ed. 1973. *The American Way of War: A History of United States Military Strategy and Policy*. Rev. ed. New York: Indiana University Press.

Wheatcroft, A., and R. Overy. 1989. *The Road to War*. London: Macmillan.

Wiarda, H. J., and H. F. Kline, eds. 1985. *Latin American Politics and Development*. London: Westview Press.

PERIODICALS, JOURNALS, AND NEWSPAPER ARTICLES

Adam, J. 1992. "Aero Space and the Military." *IEEE Spectrum: The Institute of Electrical and Electronics Engineers, Inc.*, January, 58–60.

Adams, J. R., and D. L. Crowsen. 1991. "The Army National Guard in a Changing Strategic Environment." *Military Review*, October, 35–44.

"Air Force Eyes Emerging as Weapons for Possible Use in Middle East." 1990. *Aviation Week and Space Technology*, 3 September, 36.

Alexander, J. B. 1989. "Antimaterial Technology." *Military Review*, October, 29–41.

Brewin, B. 1991. "Army Unit Constructs Mobile Drug-Trafficking Commands." *Army*, 21 October, 12.

Brzezinski, Z. 1991. "New World Order?" *School of Advanced International Studies (SAIS) Review: A Journal of International Affairs* 11(2) (Summer–Fall): 18–26.

Budiansky, S. 1990. "The Wrong Stuff: The Cold War's Legacy of Weird Weapons Lives On." *U.S. News and World Report*, 16 July, 49–51.

Budiansky, S., B. B. Auster, and P. Cary. 1991. "Preparing the Ground: U.S. Commanders Are Counting on Mobility, Air Support and New Ways to Find the Enemy." *U.S. News and World Report*, 4 February, 32–33, 39–41.

Carey, J. 1990. "If You Control Computers You Control the World." *Business Week*, 23 July, 31.

Cassani, R. 1991. "The New Economics Movement." *Development: Journal of the Society for International Development*, 84–85.

Cheney, D., U.S. Secretary of Defense. 1991. "Defending Defense: The Ever Changing Technological Arena Warrants Continued Advances in U.S. Military Forces." *Omni* 13(6) (March): 10.

———. 1992. "U.S. Defense Strategy for an Era of Uncertainty." *International Defense Review* 7.

"Cheney Fears Feud over Nuclear Arms." *International Herald Tribune*, 11 January, 5.

"Chinese Legislators Vote to Sign Nuclear Pact." *Washington Post*, 30 December, 9.

Claudio, A., 1991. "United States–Colombia extradition Treaty: Failure of a Strategy." *Military Review*, December, 69–77.

Cobb, T. W. 1992. "Neutralize Nuclear Mercenaries." *New York Times*, 2 March, 21.

Coia, D. A. 1991a. "Star Wars Paved Way for EM." *Washington Times*, 12 November, 4.

———. 1991b. "Noiseless Weapon Fires Projectiles at Hypervelocity." *Washington Times*, 28 December, 4.

Eliot, G. F. 1963. "The World Environment of 1973." *Proceedings: U.S. Naval Institute* 89(1) (January): 25–31.

"Fresh Nuclear Threat." *Daily Telegraph* (U.K.), 11 April, 12.

Glantz, D. M., 1991. "Challenges of the Future: Developing Security Issues in the Post Cold War Era." *Military Review*, December, 2–9.

Gold, P. 1992. "Pakistan Shops for Arms in Rebuff to US." *Independent* (U.K.), 13 January, 10.

Goodie, E. E. 1992. "Psychic Borderlines." *U.S. News and World Report*, 20 January, 57–59.

Greenberg, D. S. 1991. "Where Will Soviet Nuclear Warriors Seek Work?" *Baltimore Sun*, 30 December, 7.

Grimmet, R. F. 1992. "Arms Trade with the Third World: General Trends 1983–1990." *International Defense Review*, December, 55–60.

Holzer, R. 1992. "Soviet States Retain Potential to Build Large Naval Fleets." *Defense News*, 6 January, 10.

Hormats, R. D. 1991. "The Roots of American Power." *Foreign Affairs*, Summer, 132–149.

Katz, M. L. 1992. "Russian Nuke Experts Wooed." *USA Today*, 8 January, 1.

Kitfield, J. 1991. "Weapons under Fire." *Government Executive* 23(6) (June): 20–22, 24, 26.

Klein, L. R. 1992. "Turning Swords into Toasters, Again." *New York Times*, 8 January, 19.

Kutter, W. D. 1992. "Army Weapons Edge Threatened by Budget Crunch." *Army*, January, 14.

Leibstone, M. 1991. "The U.S. Defense Industrial Base: Can It Survive?" *Military Technology* 15(12): 34–38.

Lewyn, M. 1990. "Killer Computer Viruses: An Idea Whose Time Shouldn't Come." *Business Week*, 23 July, 30.

"Libya Denies Offer to Soviets." *Washington Post*, 19 February, 16.

"Magnavox Team Selected for Universal Modem." 1991. *Defense Electronics and Computing* (Supplement to IDR/12): 169.

Minasi, M. 1991. "Viruses from A to Z: Be Prepared for Any Contingency." *Compute* 13(10)(134) (October): 44–48.

Mitchell, R. 1991. "Little Star Wars: Congress Considers a Downsized SDI Based on Existing Technology." *Business Week*, 1 July, 80–81.

Morrocco, J. D. 1991. "Gulf War Boosts Prospects for High-Technology Weapons." *Aviation Week and Space Technology*, 18 March, 45–47.

Naylor, S. D. 1991. "Shoot to Disable." *Army Times*, 7 October, 26.

Nelson, J. 1992. "Arms Buildup Making Iran Top Gulf Power." *Los Angeles Times*, 7 January, 1.

"A Non-Nuclear Korean Peninsula." *Chicago Tribune*, 1 January, 20.

Norris, D. M., and E. Waples. 1992. "Information Systems and Transborder Data Flow." *Journal of Systems Management* 43(1) (January): 28–30.

Patton, J. L. 1991. "U.S. Navy Technology in the '90s." *Aerospace and Defense Science*, Spring, 58–60.

Pfaff, W. 1991. "Redefining World Power." *Foreign Affairs* 70(1): 34–48.

Powell, C. L., 1990. "U.S. Military Doctrine: The Way We Were, and Are." *Defense*, March–April, 16–20.

Reuter, T. 1991. "Pakistan Has Atomic Weapon." *Wire News*, September, 1.

Reuter, T., and T. Kalil. 1991. "Nuclear Strategy and Nuclear Winter." *World Politics* 43(4) (July): 587–607.

Robinson, L., G. Witkin, M. Cooper, and S. Minerbrook. 1991. "New Target: The Cali Cartel." *U.S. News and World Report*, 23 December, 28–30.

Ryan, R. 1992. "A Shifting Rationale by the Pentagon on Nuclear Force." *Boston Globe*, 7 January, 13.

"Saddam Helps Algeria Make Islamic Nuclear Bomb." *London Sunday Times*, 5 January, 1.

Schmemann, S. 1991. "Ukraines Bid Fleet to Swear Loyalty: Tension in Russia." *New York Times*, 5 January, 1.

Schmid, A. P. 1989. "Terrorism and the Media: The Ethics of Publicity." *Terrorism and Political Violence*, 1(4) (October): 538–564.

Schulman, S. 1991. "Artificial Germ Warfare." *Technology Review*, 18 October, 18–19.

Struck, M. 1991. "Weapons Security Chip to Get More Hill Attention." *Defense Electronics* 23(10) (October): 5.

"A Technology Czar Wouldn't Hurt." 1992. *Aviation Week and Space Technology*, 20 January, 9.

Trost, A. H., 1987. "Looking beyond the Maritime Strategy." *Proceedings: U.S. Naval Institute* 113(1) (January): 13–16.

"U.S. Navy Technology in the '90s." 1991. *Aerospace and Defense Technology*, Spring, 58, 60.

"U.S. Sabotaged Iraqi Computer, Magazine Says." *Chicago Tribune*, 12 January, 16.

Wells, S. F., Jr. 1979. "Sounding the Tocsin: NCS 68 and the Soviet Threat." *International Security* 4(2) (Fall): 116–156.

West, F. J., Jr. 1987. "The Maritime Strategy: The Next Step." *Proceedings: U.S. Naval Institute* 113(1) (January): 40–49.

White, N. D., and H. McCoubrey. 1991. "International Law and the Use of Force in the Gulf." *International Relations* 10(4) (November): 347–373.

GOVERNMENT DOCUMENTS

Bush, President G. 1991, August. *National Security Strategy of the United States*. Washington, DC: U.S. Government Printing Office.

Cheney, D., U.S. Secretary of Defense. 1991, January. *Annual Report to the President and the Congress*. Washington, DC: U.S. Government Printing Office.

Powell, C. L., Chairman of Joint Chiefs of Staff. 1992. *The National Military Strategy: 1992*. Washington, DC: U.S. Government Printing Office.

U.S. Army Armament Munitions Chemical Command. 1985. *NBC Contamination Survivability of Army Material*. Frederick, MD.

U.S. Army Training and Doctrine Command (TRADOC). 1991. *AirLand Operations: A Concept for the Evolution of AirLand Battle for the Strategic Army of the 1990s and Beyond*, TRADOC PAM 525–5. Fort Monroe, VA.

U.S. National Aeronautics and Space Command. 1977. *Laser Technology: A Survey*. Falls Church, VA: Computer Sciences Corporation.

UNPUBLISHED MATERIAL

Agarwala V. S., and L. J. Balin. 1983. "Development of Microencapsulated DNBM Quaternary Ammonium Inhibitors for Paints." *Presentation to the Electrochemical Society*. Palo Alto, CA.

Alexander, J. B. 1990. "Potential Policy Issues of Non-Lethal Defense: A Report to the Secretary of Defense, Los Alamos National Laboratory." Los Alamos National Laboratory, Los Alamos, NM.

———. 1991. "Potential Non-Lethal Policy Issues." Executive Summary. Los Alamos National Laboratory, Los Alamos, NM.

Coats, J. F. 1970, May. "Nonlethal and Nondestructive Combat in Cities Overseas." Unpublished report. Institute for Defense Analyses Science and Technology Division, Arlington, VA.

Defense Science Board, Office of the Undersecretary of Defense. 1988. "Military System Applications of Superconductors (U)." Report of the Defense Science Board Task Force. The Pentagon, Washington, DC.

Dobbs, H. H. et al. 1990, September. "Assessment of Mission Kill Concept, Requirements, and Technologies." SPC Log No. 90–1987. Defense Advanced Research Projects Agency, Arlington, VA.

Duncan, J. L. et al. 1986, April. "CARC (Chemical Agent Resistant Coating) Finishes on Laminate Armor Materials(U)." U.S. Army, Fort Belvoir Research Development and Engineering Center, Fort Belvoir, VA.

Eckhardt, W. O. et al. 1987, December 23. "Ground Based High-Power Microwave Decoy Discrimination System(U)." Hughes Research Labs, Malibu, CA.

Martin, M. D. 1991. "Non-Lethal Weapons, A Policy Planning Paper." The Pentagon, Washington DC.

Raymond, L., and A. A. Anticitil. 1989. "Hydrogen Embrittlement Susceptibility of High Strength ESR 4340 Steel Alloys for Ballistically Resistant/Tolerant Components." A Report to the Chairman of the Army Materials Technology Lab.

Standler, R. B. 1989. "Technology of Fast Spark Gaps (U)." Pennsylvania State University Park Communications and Space Sciences Lab, PA.

Sweetman, S. 1987, March. "Report on the Attorney General's Conference on Less than Lethal Weapons." U.S. Department of Justice, National Institute of Justice, Washington, DC.

U.S. Army Operations Support Directorate. 1991, June. "Army Support to National Drug Control Strategy." Unclassified message, MSG DTG: 261720Z. The Pentagon, Washington, DC.

U.S. Office of the Deputy Chief of Staff for Operations. 1991, June. "Army Counter-Drug Policy." Unclassified action memorandum. Signed by the Chief of Staff of the Army.

Wolfowitz, P. 1991, March. "Do We Need A Nonlethal Defense Initiative?" A Memorandum for the Secretary of Defense, Office of the Undersecretary of Defense. The Pentagon, Washington, DC.

Index

About the Author

DAVID A. MOREHOUSE is a third-generation officer who attained the rank of Major in the U.S. Army before resigning in 1994. He holds a master's degree in Military Art and Science from the Army's Command and General Staff College, and a doctorate in Education from LaSalle University. He writes and lectures on global peace, and is the author of *Psychic Warrior* (1996).

ISBN 0-275-95170-7

90000>

EAN

9 780275 951702

HARDCOVER BAR CODE